T0133590

Power, Thermal, Noise, and Signal Integrity Issues on Substrate/Interconnects Entanglement

Power, Thermal, Noise, and Signal Integrity Issues on Substrate/Interconnects Entanglement

Yue Ma and Christian Gontrand

CRC Press is an imprint of the
Taylor & Francis Group, an informa business

CRC Press
Taylor & Francis Group
6000 Broken Sound Parkway NW, Suite 300
Boca Raton, FL 33487-2742

Printed on acid-free paper

International Standard Book Number-13: 978-0-367-02343-0 (Hardback)

Library of Congress Cataloging-in-Publication Data

Names: Ma, Yue (Electronics engineer), author. | Gontrand, Christian, author.
Title: Power, thermal, noise and signal integrity issues on substrate/interconnects entanglement / Yue Ma and Christian Gontrand.
Description: Boca Raton : Taylor & Francis, a CRC title, part of the Taylor & Francis imprint, a member of the Taylor & Francis Group, the academic division of T&F Informa, plc, 2018.
Identifiers: LCCN 2018045497 | ISBN 9780367023430 (hardback : alk. paper)
Subjects: LCSH: Three-dimensional integrated circuits.
Classification: LCC TK7874.893 .M33 2018 | DDC 621.3815/31—dc23
LC record available at https://lccn.loc.gov/2018045497

Visit the Taylor & Francis Web site at
http://www.taylorandfrancis.com

and the CRC Press Web site at
http://www.crcpress.com

Contents

Preface

Three-dimensional integrated circuits (3D ICs) present a novel design paradigm and opportunity to address the current challenges the semiconductor industry is facing with Moore's law scaling. 3D ICs can provide potential benefits such as reduced power consumption, lower delays, and higher integration density. Through-Silicon-Vias (TSVs) are the key enablers for 3D stacking by allowing direct and less-resistive signal paths between vertically stacked circuit layers. TSVs can help reduce the wire lengths from 2D ICs and also allow replacement of chip-to-chip interconnections with intra-chip connections. Such advancements have led to design and implementation of heterogeneous systems on the same platform, that is, Flash, DRAM, SRAM-placed atop logic devices, and microprocessor cores.

However, TSV parasites contribute to power consumption and thermal integrity of the full systems compared to 2D systems.

Fundamentally, 3D ICs change how circuits are designed, analyzed, and verified. Physical design is currently facing challenges as new constraints and cost functions become more important and conventional extensions of 2D design approaches are simply not sufficient to solve these drawbacks. As circuits become more and more complex, especially for 3D ICs, new insights have to be developed in many domains: electrical, thermal, noise, interconnects, and parasites. Now, more than that, it is the entanglement of such domains that define the very key challenge, as we enter into 3D nano-electronics.

This book aims to develop this new paradigm, going to a synthesis among many technical aspects. Content is not always exhaustive, but, via some originalities, it puts new bridges over several domains of interest.

Although most of the technical specialties have developed well, some are not adapting; for example, perturbations of digital parts on very sensitive analogical systems (e.g., VCO) and new advances on electronic background noise have not been made in decades. As we go into the nano-world, we need, theoretically and experimentally, to progress to fast applicable solutions: for instance, noise sources, although autonomous, can correlate themselves.

As we know, the development of the embarked systems is subjected to the consequences of the increasingly advanced integration. These consequences appear by architectures (multi-cores) more and more complex; they present vulnerabilities of safety and are not easily testable (cf. System One Chip). A protocol is needed to reduce the risks for the possible dysfunctions of the systems embarked in radiative environment (cf. EMI).

MATLAB® is a registered trademark of The MathWorks, Inc. For product information, please contact:

The MathWorks, Inc.
3 Apple Hill Drive
Natick, MA 01760-2098 USA
Tel: 508-647-7000
Fax: 508-647-7001
E-mail: info@mathworks.com
Web: www.mathworks.com

Acknowledgments

The authors thank Fengyuan Sun, Quentin Struss, Olivier Valorge, José-Cruz Nunez-Perez, José Ricardo Cardenas, Jean-Etienne Lorival, Francis Calmon, Jacques Verdier, Pierre-Jean Viverge, Jean-Philippe Colonna, Perceval Coudrain, Louis-Michel Collin, Luc Frechette, Mahmood Reza Salim Shirazy, and Abdelkader Souifi whose valuable comments and detailed observations helped improve this book.

This work is supported by ***UpM*** *(Union pour la Méditerranée)*
and by the ***CSC*** *(China Scholarship Council).*

Authors

Yue Ma received his Electrical Engineering degree from the Ecole Centrale de Pékin University and Master's degree in computer science from the Beihang University. He obtained his Ph.D. degree in the field of microelectronics: first and second order electro-thermal parameters for 3D circuits from the Institute des Nanotechnologies de Lyon (INL) at the University of Lyon, INSA, France. His research interests include mathematics modeling, integrated circuits and systems, and computer-aided IC design with theoretical and practical issues in numerical simulation methods, applied especially to 3D ICs.

Christian Gontrand was born in Montpellier, France, and earned his M.S., Ph.D., and "State Doctorat" (Habilitation Diploma) degrees in electronics in 1977, 1982, and 1987, respectively, from the Université des Sciences et Techniques du Languedoc, Montpellier.

He worked with the Thomson "Laboratoire Central de Recherche (LCR)", Orsay from 1982 to 1984. His research interests include theoretical (electrical transport) and experimental (noise) of microwave devices (TEGFETs/HEMTs).

He joined the Laboratoire de Physique de la Matière (LPM/INSA), Villeurbanne, as a research assistant professor in 1988. He was the technical charge of the new "Centre de Microélectronique de la Région Lyonnaise" (CIMIRLY) from 1988 to 1996 and, in collaboration with the Centre National des Etudes en Telecommunication (CNET), Meylan, France, worked on new RF-compatible silicon devices.

He was the professor in semiconductor devices and circuit heading the team "Smart System Integration" at the "Centre de Génie Electrique de Lyon" (CEGELY/AMPERE) from 1997 to 2001. He has been at the head of the axis "Radiofrequency Devices, Circuits and Systems" of DE team of the Lyon Institute of Nanotechnology since 2002, dealing with noises or parasitic disturbances in mixed complex 2D and 3D RF circuits and systems.

1

General Introduction

As we know from "von Neumann bottleneck," data move between the separate MPU (microprocessor unit) and memory. In this case, latency is unavoidable. These years, processor speeds have increased significantly to about 4.2 GHz. But the memory improves mostly in the density: that is to say that the memory can store more data in a fixed space, rather than in the augmentation of transfer rates. As MPU working frequency has increased, the processor has spent an increasing amount of waiting time in order to write and read from memory. Therefore, no matter how fast the MPU can work, it is limited by the memory transfer rate [1,2].

Therefore, the next generation of AI architectures necessitates more densely integrated memory and processor. For example, the memory occupies a very big space in the IBM's TrueNorth neural network chip, Google TensorFlow, and the new generation of GPUs used in automatic driving cars has huge memories (Figure 1.1).

1.1 From Two-Dimension (2D) IC to Three-Dimension (3D) IC

Figure 1.2a illustrates that a traditional von Neumann computer system is the heterogeneous integration of logic cells (yellow), memory chips (orange), and the interconnections (gray). For example, here Data A saves the data from RF electronics and Data B saves data from camera sensors; the dynamic random-access memory (DRAM) process has a 10× higher memory density than the same generation's logic cell. Therefore, the cache in the MPU is too small for the implementation of memory, and the data must inevitably be moved through the high-energy loss route between logic and memory, twice per communication (as shown in the green line).

Figure 1.2b shows the equivalent data-transmission route for one merge stage of a bitonic sort. The light yellow logic layer on top (with black points) shows the merging of four stages of bitonic sort. When applying the structure in Figure 1.2b, the comparisons and exchange can be placed on the logic layer. Data to be calculated (Data A) and a place of storage (Data B) are located in the bottom orange memory and connected through short wires crossing the large 2D plane, rather than a border.

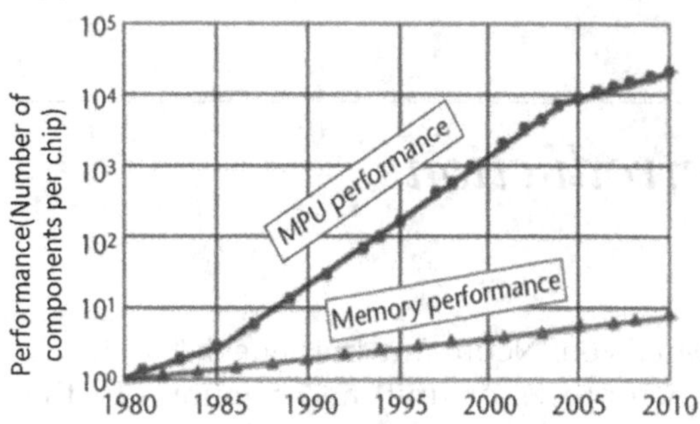

FIGURE 1.1
Performance (speed) gap between MPU and memory increasing with time [3,4] (performance's unit is number of transistors per unit cm^2).

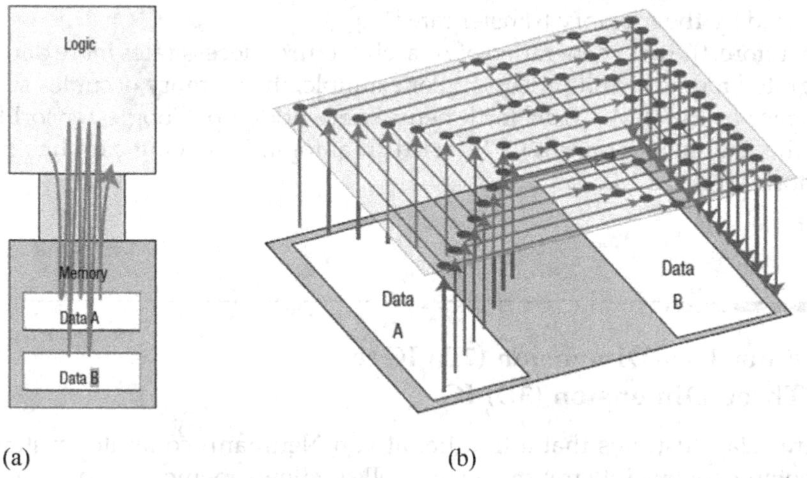

FIGURE 1.2
Advantage of 3D for interconnections. (a) 2D systems comprise logic and memory chips, with the green curve illustrating a mixed logic–memory calculation that's inefficient precisely because of this logic and memory partitioning. (b) 3D systems with tight coupling between logic and memory avoid high latency paths, bandwidth bottlenecks, and conversion of signals to high energy levels for off-chip interconnects. The blue curve shows a representative data-movement step in sorting.

For the 3D structure in Figure 1.2b, data can be read all at once from memory, and then compared and exchanged in hardware. A traditional computer will need to transport all the data from memory to the processor sequentially, so the memory bottleneck becomes a principal issue when processing a huge amount of data and other low-level kernels in artificial intelligence (AI). Therefore, the 3D circuits have a better advantage.

Moore's Law seems to be showing its limits. Because of the physics limits, the technical challenges of continuously miniaturization are difficult to overcome, which is increasing the cost and slowing the pace of scaling. In addition, the benefits of scaling are decreased because of the mismatch between the scaling and the former performance advantages. Therefore, the future semiconductor industry focuses on two technical issues. Functional diversification is the first direction. It is the main driving force of System-In-Package (SiP) architecture. The second one is 3D IC, as shown in Figure 1.3. This can gain the functionalities of one given node and also increases performance and reduces power supply. There are several driving factors for 3D integration. The benefits include better performance, reduced power supply, reduced latency, lower cost, and smaller size compared to 2.5D and 2D plane packaging. Some of the basic driving factors for 3D integration are presented in Figure 1.4 [4].

3D IC leads to a revolution of potential new scale-up. The interconnections between logic and memory have restricted systems for decades. Solving this bottleneck could continue improving system performance and advancing semiconductor industry. However, in order to get the advances, it needs some more time to work through the engineering processes (as shown in Figures 1.4 and 1.5). But every little breakthrough in the engineering processes will reduce the von Neumann bottleneck's impact. For applications that could benefit from denser integration; this could offer continuous performance increases as linewidth scaling.

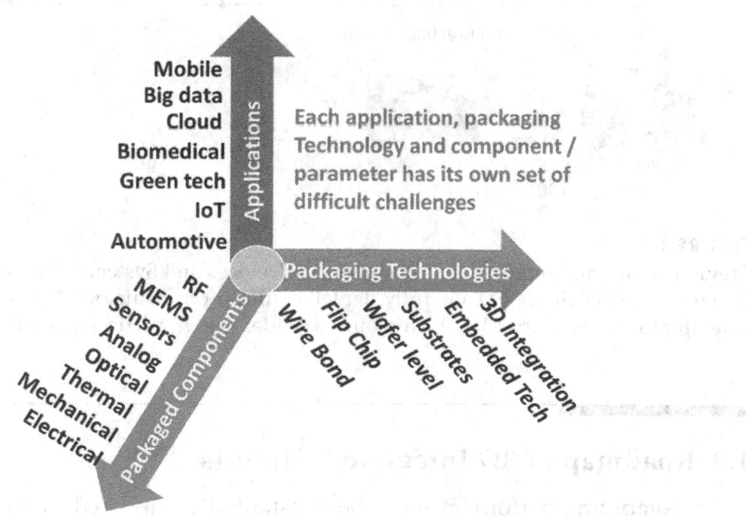

FIGURE 1.3
Elements incorporated into complex SiP packages through heterogeneous integration.

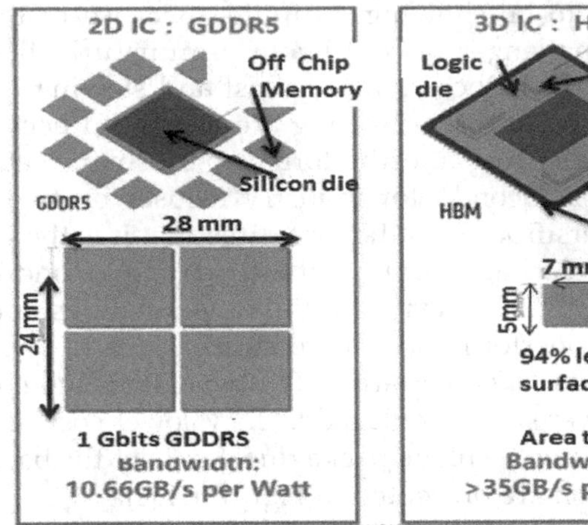

FIGURE 1.4
Driving forces for 3D Integration [6,7].

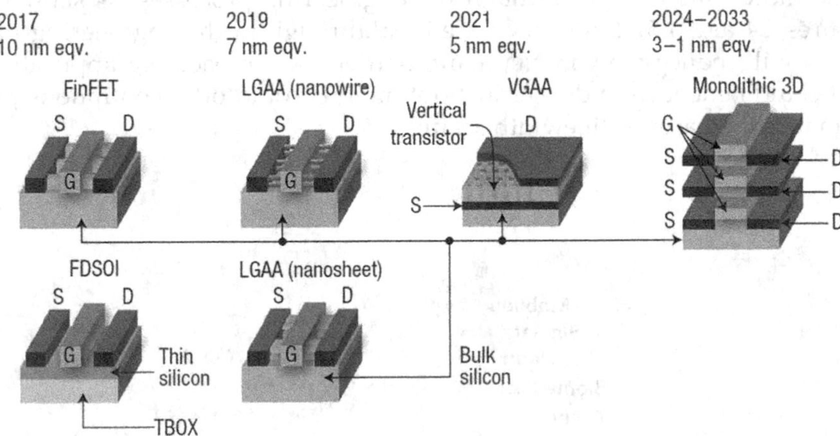

FIGURE 1.5
Preview of the 2017 International Roadmap for Devices and Systems' predictions for future 3D structures. D, drain; FDSOI, fully depleted silicon on insulator; G, gate; LGAA, lateral gate-all-around; S, source; TBOX, thin buried oxide; VGAA, vertical gate-all-around [2].

1.2 Roadmap of 3D Integrated Circuits

Computing performance can be substantially improved by monolithically integrating several new heterogeneous memory layers on top of logic layers powered by a combination of CMOS and "new switch" transistors. This new phase, called 3D power scaling, will continue to

> support an increase in transistor count at Moore's law pace well into the next decade. — *Computing in Science & Engineering* March/April 2017, vol. 19, No. 2.

The 2D integration will reach the point where no more 2D scaling will be possible; the death of Moore's law is spoken out everywhere, but the development of vertical integration of transistors and integration of layers of vertical transistors is on the way to continue increasing the transistors' density per cm^2 at a same or faster velocity than Moore's law in the future.

The 2017 IRDS (International Roadmap for Device and System) road map projects the development of integrated-circuit technology for the next 15 years (Figure 1.6).

Figure 1.5 offers a preview [2]. As we can see, 2D planar scaling in accordance with Moore's law is predicted to end until the mid-2020s, after which further advances will come mainly from the stack of layers. This miniaturization will reduce die purchase fee, energy costs, and operational benefits [8]. This evolution demonstrates that 3D VLSI offers the possibility to stack devices and enable high-density interconnects at device level (100 million "via" per mm^2), as shown in Table 1.1. 3D packaging has been available for decades in large scale, with interconnections, performance, and energy efficiency. Fully integrated 3D (3D monolithic) is supposed to be available in less than one decade.

Geometrical Scaling (1975-2003)

- Reduction of horizontal and vertical physical dimensions in conjunction with improved performance of planar transistors

Equivalent Scaling (2003-2021)

- Reduction of only horizontal dimensions in conjunction with introduction of new materials and new physical effects. New vertical structures replace the planar transistor

3D Power Scaling (2021-203X)

- Transition to complete vertical device structures. Heterogeneous integration in conjunction with reduced power consumption become the technology drivers

FIGURE 1.6
The different ages of scaling.

TABLE 1.1
Evolution of 3D Integration Options Toward 3D VLSI [1]

Options	Links	Bandwidth	Latency	Power	Time Frame
Wire-bond stack	100 s	Low	High	High	Available for 30 years
Through-silicon via (TSV) or microbump stack	1,000 s	Medium	Medium	Medium	Available for 10 years
3D VLSI stack	100,000 s	High	Low	Low	<10 years

In the 1970s, two fundamental laws became the footstones for the whole industry: Moore's and Dennard's scaling laws (cf. Figure 1.7). Dennard's scaling law presented how to improve transistor performance by means of geometrical scaling. At the beginning of the previous decade, physical scaling (the gate oxide thickness limit) came to a halt, and the advent of new transistor by reducing the horizontal dimensions and introducing new materials and new physical effects continues the Dennard's law. The new era is name equivalent scaling. And the vertical structures begin replace planar transistor. The predictive revolution date is 2021. At that moment, device built via 2D scaling will reach a fundamental manufacturing 2D physics limit; the transition to complete vertical device structures, heterogeneous integration in conjunction with reduced power consumption, becomes the technology drivers [2,3,4].

The IRDS (IEEE *International* Roadmap for Devices and Systems) projects also five steps of functional transition from 2D to 3D VLSI: (here include the functional heterogeneous devices as sensors, new storage devices, alternative logic, and memory devices or circuits)

- Two stacked layers with large-scale components such as analog, I/O, and power management, MEMS in one layer and high-performance logic and memory in the other.
- Monolithic integration of two layers, where each layer contains one of the two fundamental transistor types in logic, NMOS and PMOS (n-channel and p-channel MOS), stacked on top of each other in order to increase logic density.
- Two layers: logic and memory.
- Analog, I/O, and RF connectivity as an extra layer, giving more freedom to include special devices in the design.
- 3D VLSI with fine-pitch logic-on-logic as well as special functional layers exploiting new architectures.

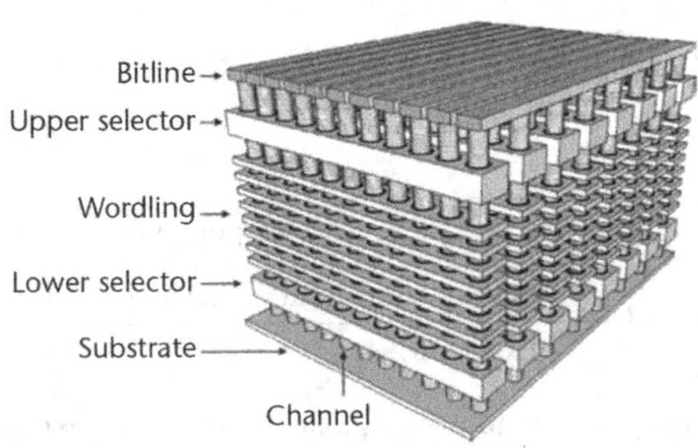

FIGURE 1.7
Example 3D architecture of future monolithic integrated circuits [3,4].

1.3 Development of 3D IC

Today, memory producers are facing similar problems because they have no space in 2D, and the cost of producing memory circuits of small dimensions keeps rising while the number of stored electrons in the floating gate keeps decreasing. To solve these problems, memory producers have already worked out and announced several new architectures that stack multiple layers of memory on top of each other in a single IC. As many as 48 and 96 layers of flash memory have been reported. Table 1.2 shows the prediction of flash memory devices for more than 100 layers. Therefore, heat removal and reduction of power will be the two main issues in the 3D power scaling.

Beyond 2020, the future ICs architecture will change to a type of structure similar to the one in Figure 1.7. This new method of scaling, which is

TABLE 1.2
Flash Memory Trends

NAND Flash							
Year of Production	**2015**	**2016**	**2020**	**2022**	**2024**	**2028**	**2030**
2D NAND flash uncontacted poly ½ pitch-F (nm)	15	14	12	12	12	12	12
3D NAND minimum array ½ pitch-F (nm)	80nm	80nm	80nm	80nm	80nm	80nm	80nm
Number of word lines in one 3D NAND string	32	32–48	696	96–128	128–192	256–384	38512
Domain cell type (FG, CT, 3D, etc.)	FG/CT/3D	FG/CT/3D	FG/CT/3D	FG/CT/3D	FG/CT/3D	FG/CT/3D	FG/CT/3D
Product highest density (2D or 3D)	256G	384G	768G	1T	1.5T	3T	4T
3D NAND number of memory layers	32	32–48	696	96–128	128–192	256–384	38512
Maximum number of bits per cell for 2D NAND	3	3	3	3	3	3	3
Maximum number of bits per cell for 3D NAND	3	3	3	3	3	3	3

Source: www.itrs2.net

capable of stacking multilayers in a 3D devices level, can substantially accelerate Moore's law. Research on new switches began in 2005. By 2010, the tunnel FET (TFET) seems a potential replacer of MOS as shown in Figure 1.8, because of its lower operating power [9]. Negative capacitance FET (NC FET) has also shown as another lower operating power device than MOS [10,11]. CNTs (carbon nanotube) have become the most viable switch later in the next decade, because of its energy saving and performance improvement [12]. The number of transistors that can be integrated on a cm^2 of silicon will continue to increase at Moore's law pace for the next 10 or more years by means of 3D integration. Till now, transistor speed has continued to increase during this decade, and the operation frequency of transistor has already arrived at 100 GHz if power dissipation was not a bottleneck, but because of the power dissipation limits of ICs; today's MPU's operating frequency has been limited to a few GHz (CPU intel Core i9 4.2 GHz). Therefore, in the next decade, the goal of IC design has shifted from high frequency design to low power and better heat dissipation design.

Three-Dimensional (3D) Integration and Packaging has been successful in mainstream devices to increase logic density and to reduce data-movement distances. It solves the fundamental limits of scaling (cf. Figure 1.9), for example, increasing delay in interconnections [14], development costs and variability [15]. Most memory devices shipped today have some form of chip-stacking involved. 3D integration has different technologies; with the development, they are wire bonding and flip chip for SiP, through-silicon via

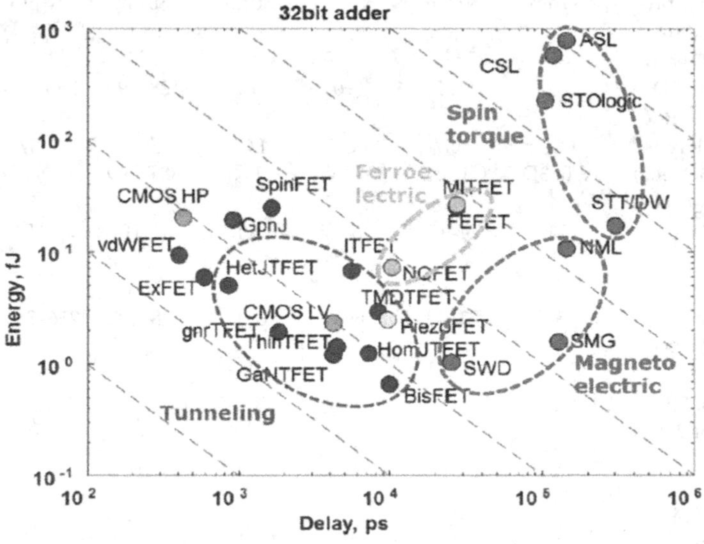

FIGURE 1.8
Switching energy vs. delay of a 32-bit adder [13].

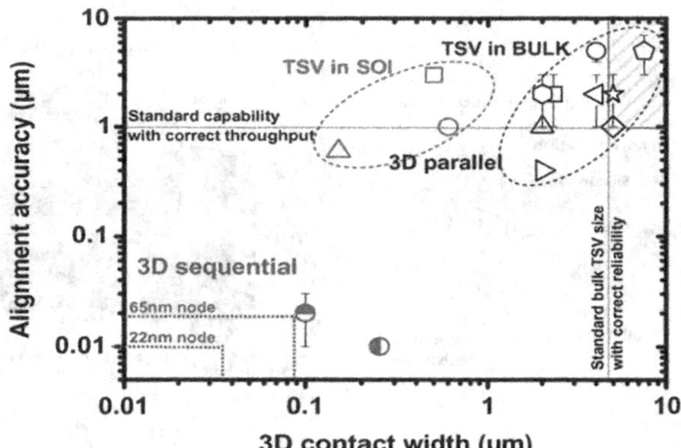

FIGURE 1.9
Alignment capability versus 3D contact width in parallel and sequential integration schemes [3,4].

(TSV) in bulk, TSV in SOI (silicon on insulator), 3D parallel, 3D sequential (MIV: monolithic inter via).

In the bulk level design (>2 µm), there have two stacked dies types: the interconnection via substrate and direct connection between dice. The typical interconnection via substrate technology is wire-bonding stack. The wire-bonding stack, as shown in Figure 1.10a, has a low-density, low-bandwidth, high-latency, and high-power dissipation characteristic (the exact numbers can be found in Table 1.1). TSV in bulk is the typical one of the direct connection between dice, which is the main stream of 3D interconnection because that they do not need substantial change to the existing fabrication flow, as shown in Figure 1.10b. The size in CMOS image sensors using "via last" approach has a diameter of about 50 µm [3,10,16–19]; it has smaller size for 3D stacked memory (NAND, DRAM …) and logic circuits (MPUs with cache memory). Stacking involves drilling holes through-silicon chips to provide electrical connections between layers of stacked silicon chips. Chip stacks up to 40 layers deep are available that have lower engineering costs than adding layers lithographically or epitaxial deposition. TSV offers lower bandwidth and less efficient connectivity between chip layers than adding layers with photolithography processes, but far higher bandwidth and efficiency than using conventional chip packaging. It has an advantage of 10X more links in the circuit and better bandwidth, better latency, and lower power supply [17,18,20–22]. However, the keep-out zone (KOZ) [3] required for TSVs and limitations on the die alignment precision generate limits on the device integration density that can be designed using TSV-based 3D stacking [2].

Below 1 µm, in this field, the 3D technology's interest stays at the transistor scale design. Figure 1.9 shows that the alignment accuracy between the stacked transistor layers with regard to the 3D contact width (pitch), the 3D parallel integration [2] as shown in Figure 1.10(c), enables the continuously

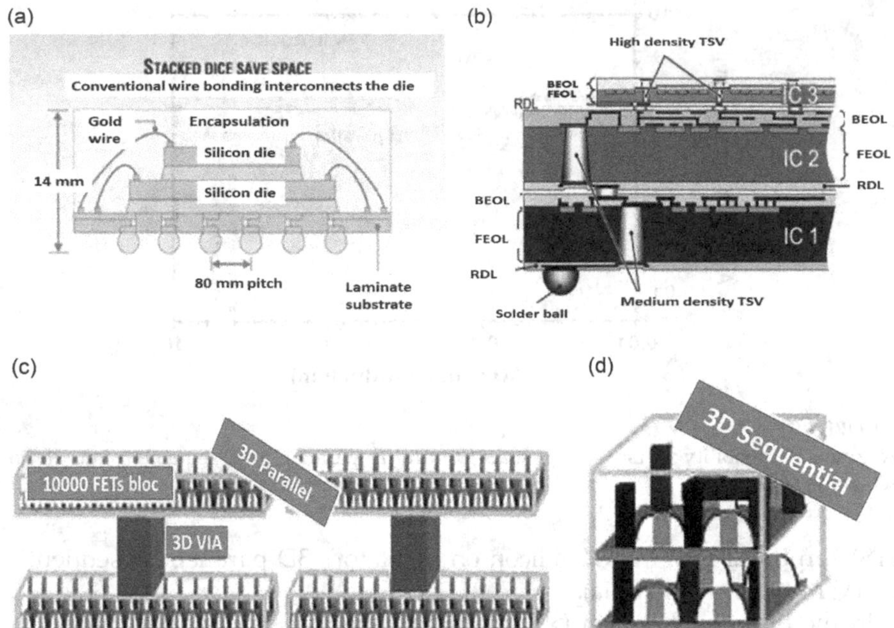

FIGURE 1.10
Different 3D interconnection technologies. (a) Wire bonding. (b) Through-silicon via (high-density TSV & medium-density TSV). (c) 3D parallel integration. (d) 3D sequential integration.

reduction after 3D TSV; the pitch reduction has been reduced to 1 µm. Although some 3D SOI architectures can reduce the 3D contact size, the 3D contact pitch remains however limited by the alignment performance, which is limited to about 1 µm, to guarantee a correct throughput.

3D sequential integration is the next generation of 3D integration technologies; it's also named monolithic 3D (M3D) [1–4]. It has been demonstrated that 3D interconnections and planar contact pitches are matching even for advanced nodes [2–4] and it has the potential to achieve nano-scale device integration compared to TSV-based 3D stacking [9], and for the 3D parallel integration, one via has a connection of a few thousand transistors blocks [3,4], as shown in Table 1.3. That's too coarse-grained compared with monolithic

TABLE 1.3
Ratio between 3D via and Planar Contact Pitch in Sequential and Parallel Case [12]

	Parallel		Sequential
Pitch	0.5 µm	1 µm	10 µm
Dvia3D/DCONT, 45 nm	1/1000	1/10000	1/1

inter-tier vias (MIVs). Hence, that's why the 3D sequential integration is the only possibility to fully exploit the potential of the third dimension, especially at the transistor scale. There are three categories of M3D integration depending on the pitch of tier partitioning: transistor-level, gate-level, and block-level M3D integration. In transistor-level M3D, P-channel transistors are placed on one tier and N-channel transistors on another tier, and connection between them is established by MIVs. In gate-level M3D, standard cells are split across multiple tiers, and inter-cell connections which cross tiers utilize MIVs. Block-level M3D design, which is the coarsest-grained design style, partitions functional blocks, and inter-block MIVs are deployed to deliver signals between functional modules in different tiers. Unfortunately, nowadays the commercial EDA tools to support 3D integration are temporally not available. Therefore, various studies have been reported on M3D IC design using 3D commercial tools. M3D has also the disadvantage of device and interconnect performance mismatch between tiers, testing challenges, cost, etc.

Implementation of a 3D structure including CNFETs, resistive RAM (RRAM), and spin-transfer torque magnetic RAM (STT-MRAM), has already been proposed and analyzed by using N3XT ("N": next, and "3": three event

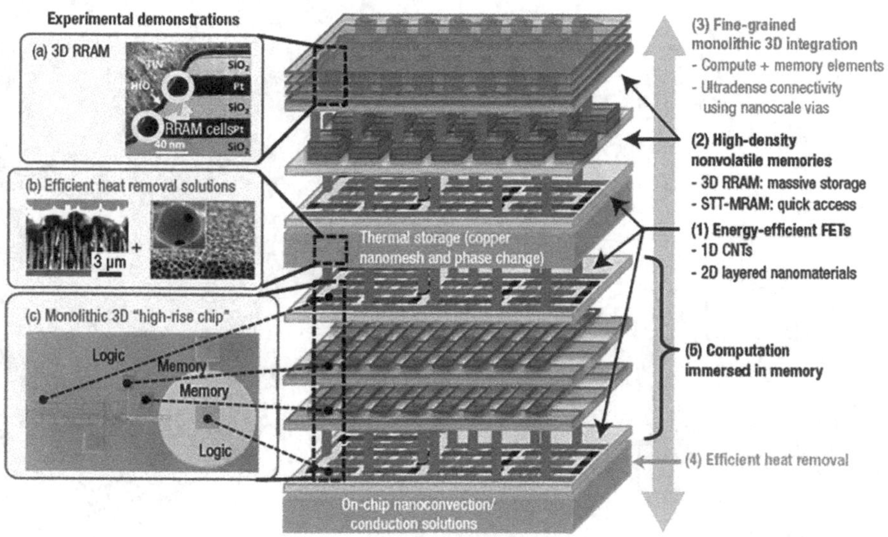

FIGURE 1.11
Monolithically integrated 3D system enabled by Nano-Engineered Computing Systems Technology (N3XT). On the right are the five key N3XT components. On the left are images of experimental technology demonstrations: (a) Transmission electron microscopy (TEM) of a 3D resistive RAM (RRAM) for massive storage. (b) Scanning electron microscopy (SEM) of nanostructured materials for efficient heat removal (left: microscale capillary advection; right: copper nano-mesh with phase-change thermal storage). (c) SEM of a monolithic 3D chip for high-performance and energy-efficient computation. CNTs, carbon nanotubes; FETs, field-effect transistors; STT-MRAM, spin-transfer torque magnetic RAM [2].

pillars of *ideas, innovation,* and *inspiration*)—Nano-Engineered Computing Systems Technology_proposed by University of Stanford and University of California Berkeley [21]. The structure is shown in Figure 1.11. The N3XT model has been demonstrated to have a 23X speedup and 37X energy reduction; that's to say, it will have a total energy-delay product benefit of 851 [3,4].

2

Substrate Noise in Mixed-Signal ICs in a Silicon Process

2.1 Introduction

Substrate coupling in mixed-signal integrated circuits (ICs) and especially in wireless communications systems-on-chip (SoCs) creates important performance degradation of the analog/RFMS (RF and mixed signal) circuits integrated on the same die as large digital systems. Couplings occur between noise transmitters, which, in most cases, are fast switching digital blocks, and a noise receiver, which, generally, is a sensitive analog RF block or mixed-signal architecture and takes place due to the capacitive and resistive nature of the substrate-interconnects and substrate-devices interfaces. The very dense integration to minimize the product area to cost ratio, and the new design concept of making analog/RF operations using digital architectures, renders substrate crosstalk as one of the most crucial obstacles. Especially it is the case of technology multi-sourcing (fabricating the same silicon product footprint in different CMOS technologies, but with the same minimum MOSFET transistor length, so as to minimize the cost between the silicon process vendors), and, in particular, in mobile communication SoC design. This becomes even more severe due to complete radio spectrum usage from 3 kHz to 30 GHz and specifications with signal/power below background noise. Moreover, this will become a blocking point in reaching millimeter wave IC design and moving from the 4G/LTE mobile communications to the 5G mobile communications design era, and from 28 nm and 20 nm planar CMOS processes, to 14 nm and 10 nm FINFET [3].

Many papers and research programs have been published [2,20–22] in recent years, but this topic has still to be investigated, as it is a complex problem to quantify with a clear method and easy-to-use tools. As an example, the accurate substrate modeling by means of advanced finite element methods with powerful simulators is time wasting without taking into account the chip in its application (package, printed circuit board—PCB, etc.). In a pragmatic way, the goal of this chapter is to present simply the problem, and propose some basic rules associated with a methodology and some tools to quantify and reduce substrate noise in future designs.

The context "silicon technology" is based on low- or high-resistivity substrate, the substrate noise generation, and propagation. In the rest of the paper, noise generation will be simplified by digital power-supply ringing transmitted to P-substrate or N-wells. A methodology based on the IC Emission Model—ICEM standard model [23–27]—is applied, with an extension to substrate coupling. Associated tools are presented to quantify the voltage supply ringing, the substrate equivalent resistance, and the analog ground fluctuations. A comparative study of different substrate isolation techniques applied to a virtual IC to determine the best design strategies is made, with an illustration of the methodology applied to a mixed-signal IC integrating different digital blocks and a voltage-controlled oscillator in a BiCMOS SiGe process.

2.2 Ground and Substrate Noise Mechanisms

One of the main phenomena that induce substrate noise in a mixed-signal circuit is the power and ground supply voltage fluctuations that are transmitted into the substrate through all substrate biasing contacts [23–28]. The power and ground supply lines are not perfect and introduce several parasitic elements: resistances, inductances, and capacitances. In a large digital circuit, high peaks and fast slew rate on supply current create power-supply noise in the supply network due to *RLC* network formed by all these parasitic elements at different levels of an SoC: PCB, package, and the circuit itself. Other mechanisms create substrate disturbances in mixed-signal devices [29]. These mechanisms are just referred here and shown in Figure 2.1 which represents a single digital inverter.

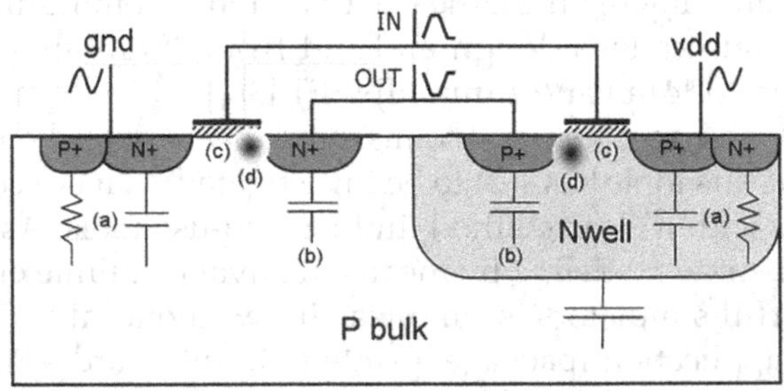

FIGURE 2.1
Substrate noise injection mechanisms in a digital inverter.

The main phenomenon of substrate coupling is briefly described: it is the power-supply noise (Figure 2.1a). The signal transitions can also be coupled to the substrate through different physical structures: the input MOS capacitance (Figure 2.1c) of all transistors, the output drain (Figure 2.1b) to substrate capacitances, and the metal interconnection to substrate capacitances. Other phenomena can induce parasitic substrate currents like impact ionization (Figure 2.1d), photon-induced current, and diode leakage current.

2.3 Substrate Noise Propagation in Low- or High-Resistivity Silicon Substrate

After the substrate noise generation described in the previous section, the propagation has to be quantified through the substrate. Silicon material is characterized by a cutting frequency $f_c = \sigma/2\pi\varepsilon$, where σ is the conductivity and ε is the permittivity. With high-resistivity substrate (>10 Ω cm), silicon can be considered as purely ohmic for signal frequency below 10 GHz [12]. Then the silicon equivalent transfer function can be defined as an attenuation factor that can be calculated with point-to-point equivalent resistance. At this step, one has to consider the nature of the substrate: low- or high-resistivity silicon substrate.

Considering low-resistivity substrate, the substrate model depends on the comparison between the epitaxial layer thickness (W_{epi}) and the distance D between substrate taps. As shown in Figures 2.2 and 2.3 (case b); when the distance is up to $4 \times W_{epi}$, all the current flowlines pass through the low-resistivity substrate that can be considered as a unique point. The resistance variation function of the distance D is represented in Figure 2.4 (substrate tap area is 20×20 μm^2, $W_{epi} = 5$ μm @ 10 Ω cm, substrate thickness is 300 μm @ 0.05 Ω cm).

For high-resistivity substrate, the die backside connection is first of interest. If the die backside is floating, the resistance between two taps increases with the distance (Figure 2.4). If the die backside is correctly connected to the

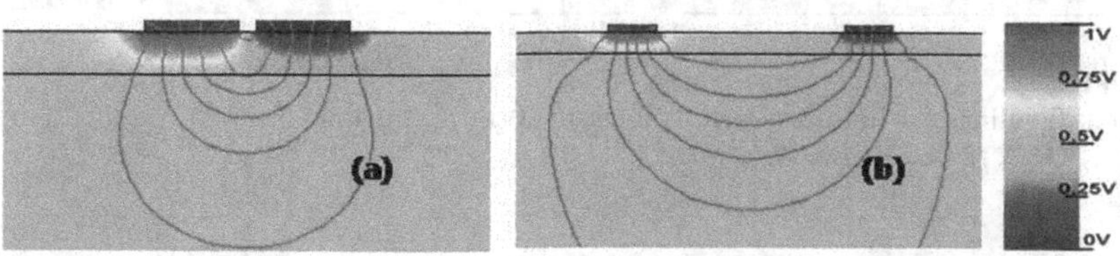

FIGURE 2.2
Flowlines for low-resistivity substrate with epitaxial layer (W_{epi} thickness): (a) $D < 4 \times W_{epi}$ and (b) $D > 4 \times W_{epi}$.

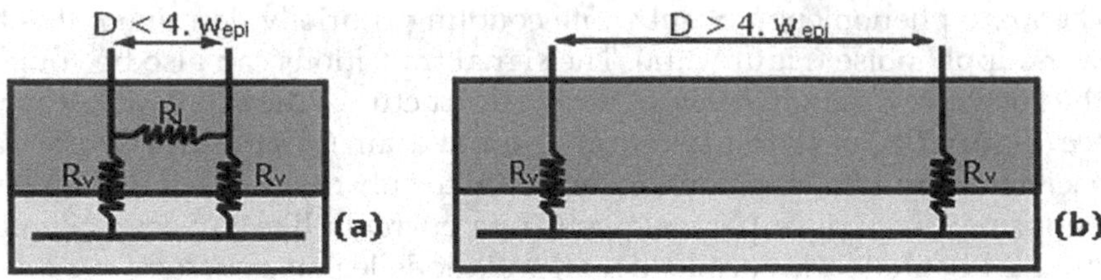

FIGURE 2.3
Equivalent model for low-resistivity substrate with epitaxial layer (W_{epi} thickness): (a) $D < 4 \times W_{epi}$ and (b) $D > 4 \times W_{epi}$.

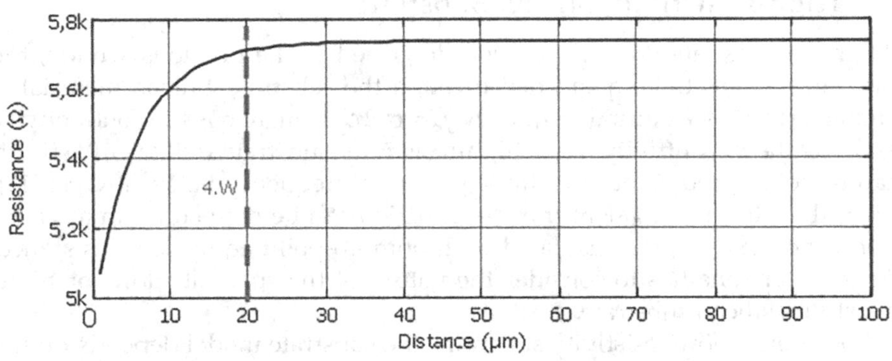

FIGURE 2.4
Equivalent resistance function of the distance for low-resistivity substrate with epitaxial layer (W_{epi} thickness).

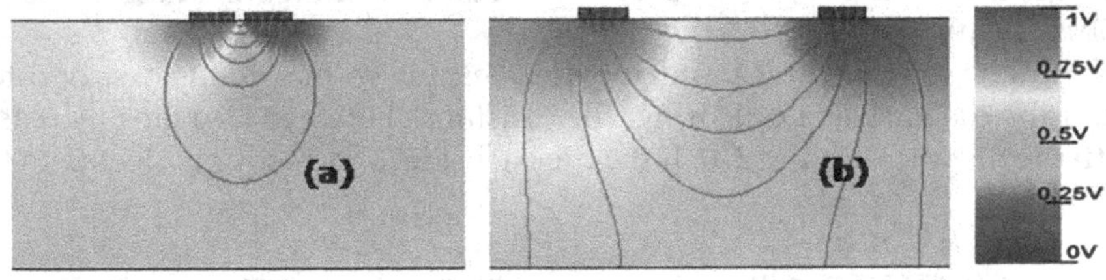

FIGURE 2.5
Current flowlines for high-resistivity substrate ($W_{substrate}$ thickness) with grounded backside: (a) $D < W_{substrate}/2$ and (b) $D > W_{substrate}/2$.

ground, the resistance depends on the distance between the taps compared to the substrate thickness $W_{substrate}$ as illustrated in Figures 2.5a–2.7 (substrate tap area is $20 \times 20\ \mu m^2$, $W_{substrate} = 50\ \mu m$ @ 6.7 Ω cm).

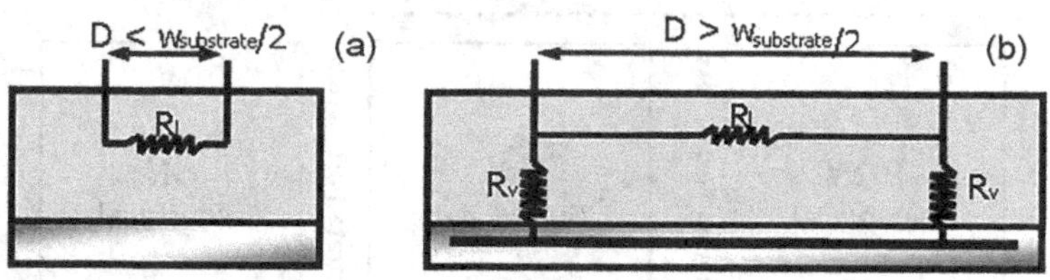

FIGURE 2.6
Equivalent model for high-resistivity substrate ($W_{substrate}$ thickness) with grounded backside: (a) $D < W_{substrate}/2$ and (b) $D > W_{substrate}/2$.

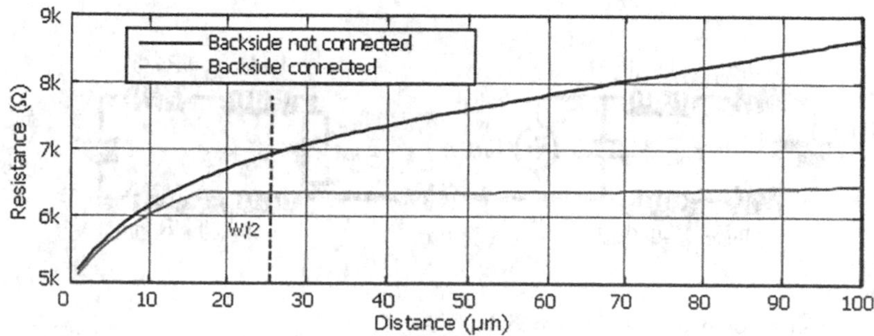

FIGURE 2.7
Equivalent resistance function of the distance for high-resistivity substrate ($W_{substrate}$ thickness) with floating or grounded backside.

2.4 Modeling Methodology

2.4.1 Ground and Substrate Noise Modeling

In this approach, we consider the power-supply noise as the unique substrate perturbation source. This method drastically simplifies the problem and gives quite good results, as shown in the following sections. In order to model power and ground voltage oscillations, we have chosen to use the ICEM (Integrated Circuit Emission Model) approach [24–28]. The classical ICEM has been extended with a Substrate Network Sub-Model [19,20]. Figure 2.8 gives an overview of the extended-ICEM model:

The Passive Distribution Network Sub-Model is supposed to model the parasitic elements of power-supply or signal lines. The Internal Activity Sub-Model characterizes the dynamic current consumption of the device. The Substrate-Network Sub-Model is expected to model the substrate propagation of parasitic signals from aggressor parts to victim parts.

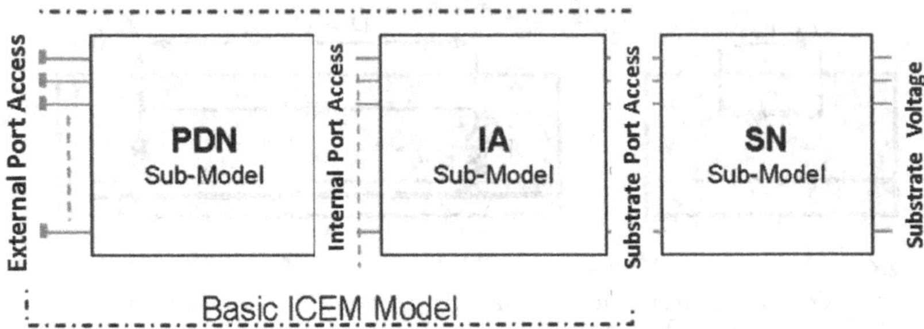

FIGURE 2.8
Conceptual architecture of extended ICEM.

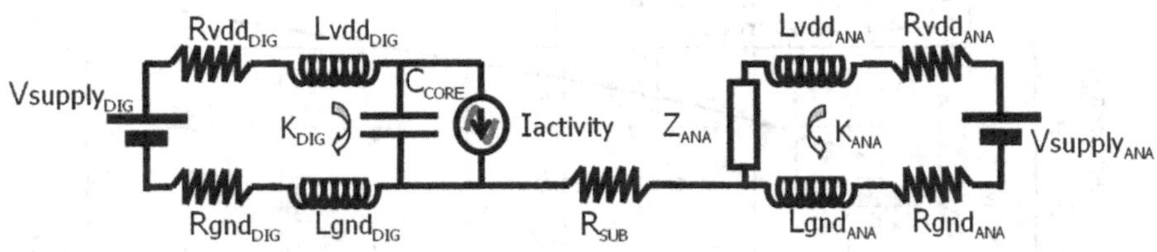

FIGURE 2.9
Basic architecture of extended-ICEM model.

2.4.2 Developed Tools

Basically, the conceptual extended-ICEM methodology can be described in its simplest form by the electrical diagram presented in Figure 2.9 (one digital block with one *Vdd-gnd* power-supply pair). Nevertheless, some parameters have to be determined for each IC project. Therefore, the authors propose to use two software applications developed in JAVA language. The first application is developed for ground and power-supply bounce effects. The second one concerns the substrate extraction. These applications can be used at any step in the design phase: from the preliminary study to the final optimization (post-layout) as the parameters can be refined in time.

2.4.2.1 Application Developed for Ground and Power-Supply Bounce Effects

The first application deals with quantifying the ground and power-supply fluctuations (bounces). This can be done in three steps. Firstly, the user describes the passive elements: R, L, C_{core}, and K. The parameters R and L are the parasitic resistance and inductance of the lines (PCB, socket, package, on-chip metal rails). K is the coupling factor (mutual inductance) between the lines (mutual between bonding wires, typical value is around 0.5). In digital ICs, the power and ground parasitic elements are mainly due to the package effects [29,30], power-grid effects can be neglected in a first order approach. Package lead and bonding wire lengths are often longer than on-silicon metal lines and induce mainly parasitic inductances. C_{core} is the equivalent digital capacitance; it depends on the technology (number of gates per mm²) and the digital area. Typical values range from 0.15 to 0.3 nF/mm² for 0.35 μm CMOS technology and from 0.8 to 1.6 nF/mm² for 90 nm. These passive parasitic elements can be estimated or determined from the Z_{11} S-parameter measurement (Figure 2.10a,b).

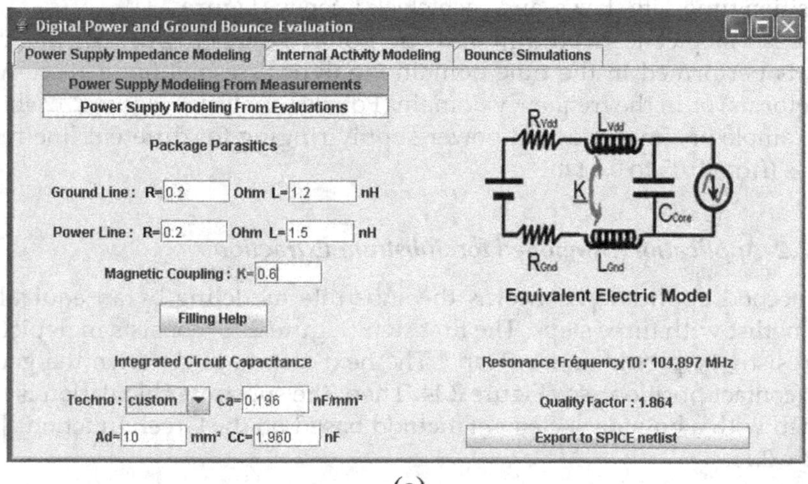

(a)

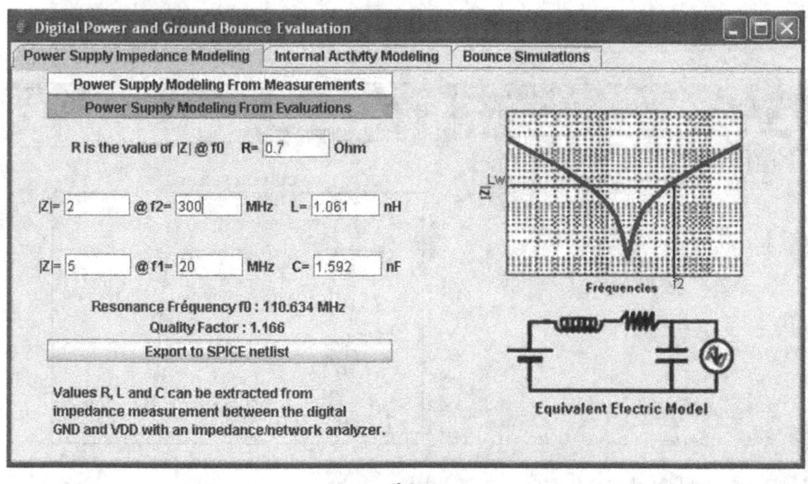

(b)

FIGURE 2.10
Dedicated application to quantify digital ground and power-supply bounces (passive distribution network).

a. Passive parasitic elements can be estimated
b. Passive parasitic elements can be determined from the Z_{11} S-parameter measurement.

In a second step, the internal activity of the digital block is described in a simple way. This current source is the superposition of periodical current pulses. The characteristics of each pulse (height, width, frequency, rising and falling time) can be easily determined from general digital circuit parameters: its average consumption, its clock frequencies, its rising and falling transition times, its skews, and each clock latency (Figure 2.11).

The last step concerns the ground and power-supply bounce calculations. This is performed in the time domain (analytical resolution of piece-wise waveforms) or in the frequency domain (Fourier analysis). Figure 2.12 shows an example of simulated Vdd power-supply ringing for different line resistances (from 0.05 to 0.5 Ω).

2.4.2.2 Application Developed for Substrate Extraction

The second application concerns the substrate modeling by an equivalent {*RC*} netlist with three steps. The first step (Figure 2.13) consists in depicting the resistivity profile under P-taps. The next step is to describe the geometry (contact position), see Figure 2.14. Then, the substrate calculation is performed with a boundary element method based on the Green functions [19]: Figure 2.15.

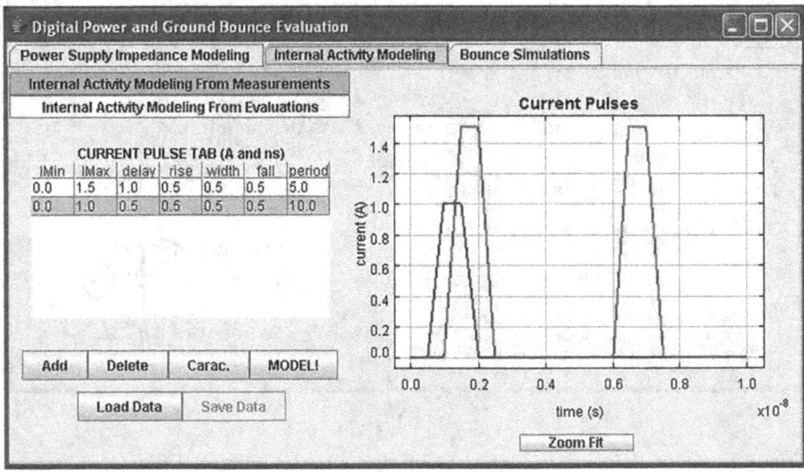

FIGURE 2.11
Dedicated application to quantify digital ground and power-supply bounces (internal activity).

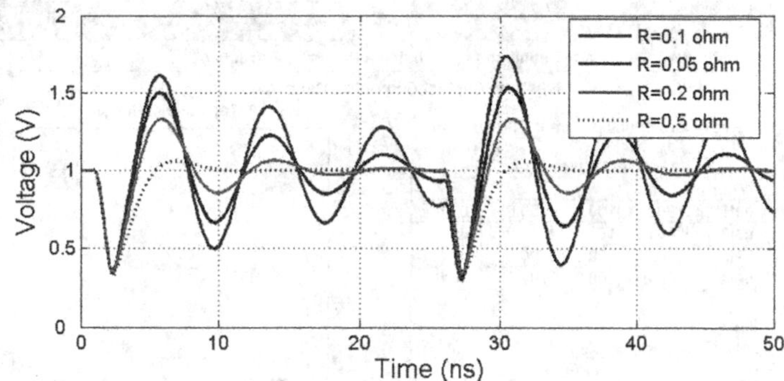

FIGURE 2.12
Simulated power-supply bounces for different line resistances.

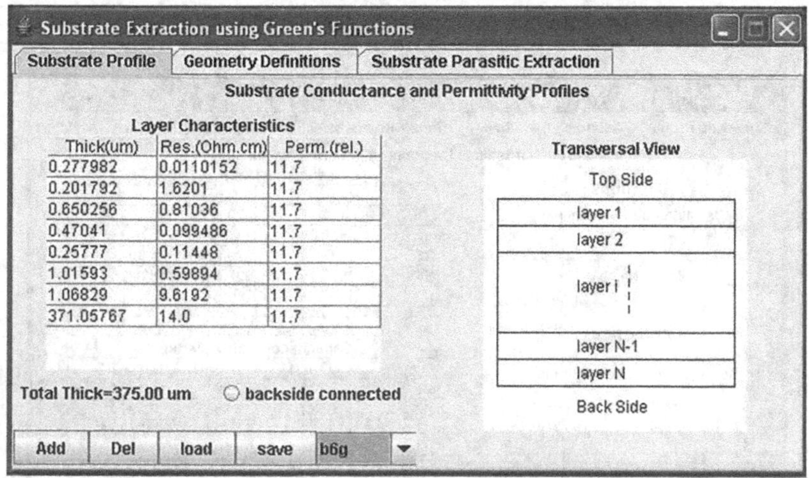

FIGURE 2.13
Dedicated application to substrate extraction (resistivity profile).

2.4.3 Basic Rules to Reduce Digital Power-Supply Network Ringing

With the presented tools, the user can estimate how the analog ground will fluctuate due to the activity of the digital(s) block(s) even if the supplies are completely separated (Figure 2.16). Moreover, these tools can help understand malfunctions of digital blocks with an unstable digital power supply.

The passive distribution network presents a resonant frequency $f_{res} = 1/2\pi\sqrt{LC_{core}}$ and a quality factor $Q = L\omega_{res}/R$. The passive distribution network will oscillate for $Q>0.5$. To suppress or reduce power-supply ringing, the designer can control the following parameters: R (line resistance from board or inside chip), L (line inductance), *and* C_{core} (core capacitance and

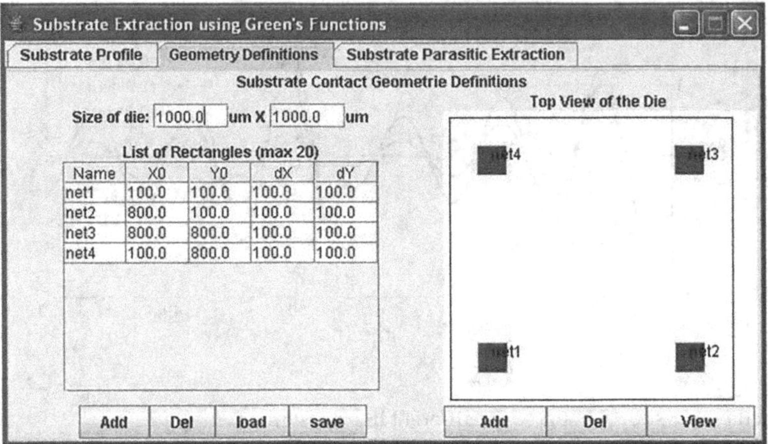

FIGURE 2.14
Dedicated application to substrate extraction (geometry definition).

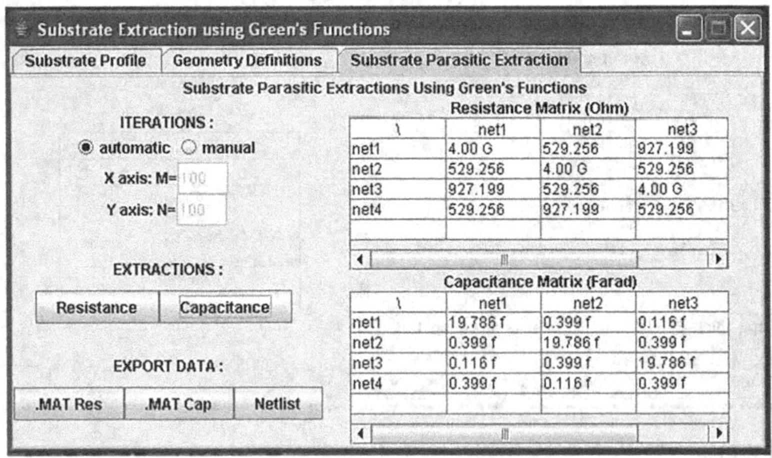

FIGURE 2.15
Dedicated application to substrate extraction ({*RC*} substrate matrix).

additionally on-chip decoupling capacitance). Hereafter, we summarize the influence of these parameters on the digital power-supply ringing:

- increasing R leads to a decrease in oscillation amplitude without any influence on the oscillation frequency (Figure 2.16),
- increasing L leads to a decrease in the oscillation frequency without any influence on the oscillation amplitude,
- increasing C_{core} leads to a simultaneous decrease in oscillation amplitude and in oscillation frequency.

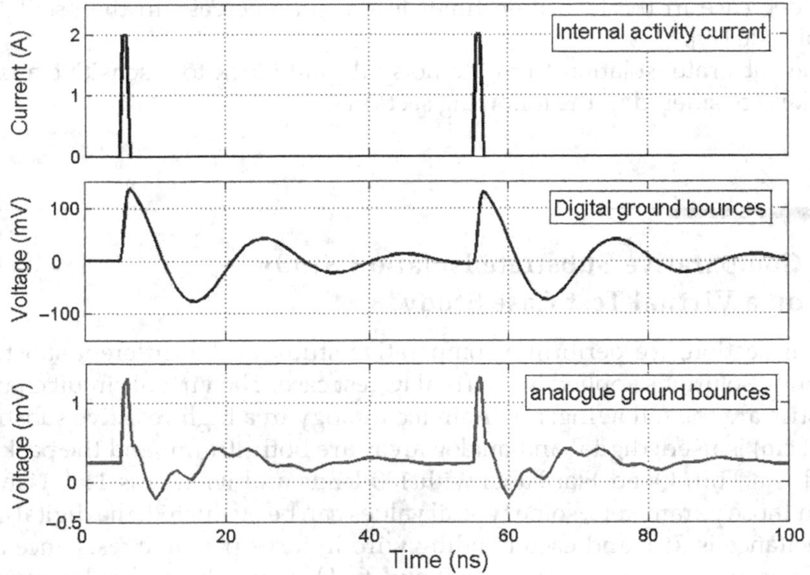

FIGURE 2.16
Simulated analog ground fluctuations.

An intuitive way to decrease power and ground bounces is to add a huge decoupling capacitance on the power network of the PCB. In fact, this external decoupling capacitor just filters the power network parasitic voltage, and is not enough to decrease significantly the interferences [24,28–34]. The main disturbing frequency is due to the ringing of the *RLC* path composed of the bonding wire inductance and resistance and by the core capacitance of the digital part of the circuit. Increasing as much as possible the off-chip capacitance will not avoid the on-chip natural ringing. Another simple solution is to add a serial resistor to the power network (on the PCB or inside the chip). The added resistor decreases the quality factor of the *RLC* network and, in the same time, dampens the induced oscillations. The obtained low-pass filter gives some good results. But, too great a resistance of the power network induces a high power consumption of the chip and also could cause some functional mistakes. These are the reasons why this kind of noise reducing technique is not recommended for a large digital circuit but can be applied to a small digital application such as a frequency synthesizer.

The internal activity (digital equivalent current source) has also a direct link with the power and ground oscillation amplitude. For clock frequency below resonant frequency, the power-supply ringing will be enhanced when the resonant frequency is a harmonic of the clock frequency (the worst case is for $f_{\text{res}} = 2f_{\text{clock}}$). For resonant frequency below clock frequency, the passive distribution network acts as a low-pass filter. An increase in the average consumed digital current leads to an increase in the oscillation amplitude.

The decrease in the transition times leads to a decrease in the oscillation amplitude [19].

The substrate isolation from the noisy digital block to a sensitive analog block is considered in the following section.

2.5 Comparative Substrate Isolation Study on a Virtual Test Case Study

In this section, we perform a comparative study of the different substrate isolation solutions applied to a virtual IC test case. The virtual circuit characteristics are the following: a 130-nm technology in a high-resistive substrate (10 Ω cm) is used; digital and analog areas are both 10 mm^2 and the package used is a Thin Quad Flat Panel with 100 leads and its area is 14 × 14 mm^2. From those parameters, some typical values can be evaluated: the digital core capacitance is 7 nF and each bonding wire induces parasitic resistance and inductance of, respectively, 120 mΩ and 6 nH. From these simple assumptions, a first order electrical model can be elaborated and can be helpful for technological and layout choices. Here is the list of solutions that can be used in order to decrease the substrate noise impact in such a device: substrate choice (high or low resistivity), die backside connection (grounded or floating), number of *Vdd-gnd* power-supply pairs, number of bondings, layout guard-ring around block, layout triple-well (Figure 2.17). Good efficiency of guard-rings or triple-wells is obtained by biasing each structure with a specific supply/ground pad. The triple-well technique would offer the better substrate isolation at low frequencies. The N-well has to be as small as possible in order to decrease the reverse biased diode area, and so the capacitance between analog or digital part and the substrate. A small capacitance can filter higher frequencies.

The simulated isolation is quantified by the transfer function between the differential analog supply and the digital internal activity in dB (from Figure 2.9). Here are the different configurations:

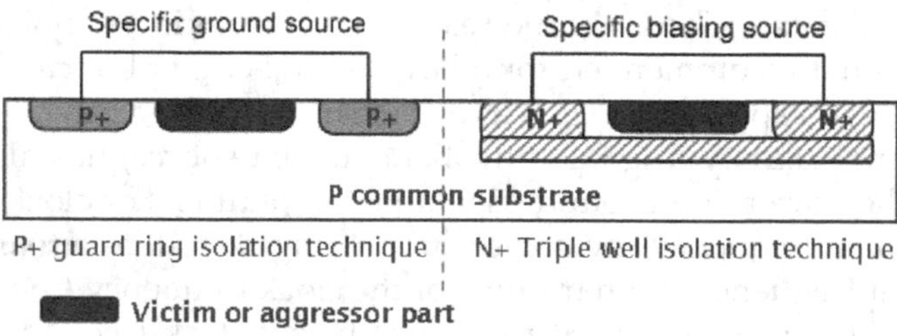

FIGURE 2.17
Layout substrate isolation techniques: guard ring and triple-well.

- 1st case is the reference: no layout protection, high-resistivity substrate with floating die backside.
- 2nd case: the die backside is grounded.
- 3rd case: high-resistivity substrate is replaced by low-resistivity substrate (floating die backside).
- 4th case: high-resistivity substrate is replaced by low-resistivity substrate (grounded die backside).
- 5th case: the noisy digital block and the sensitive analog block are surrounded by two guard-rings.
- 6th case: the sensitive analog block is isolated in a triple well.
- 7th case: the noisy digital block is isolated in a triple well.

The different transfer functions are represented in Figure 2.18. On these different waveforms, we observe a first resonance around 50 MHz; this corresponds to the digital supply network resonance frequency. The second resonance peak is around 200 MHz and corresponds to the analog supply network resonance frequency.

For this virtual IC, the best isolation technique to reduce substrate coupling is to surround the digital and the analog block with two guard-rings. The *worst* solution is to use a low resistive substrate with the floating backside. This result cannot be generalized to other designs as each case is specific. Electromagnetic coupling between bonding wires is also not taken into account in this simple virtual study. This explains the excellent isolation

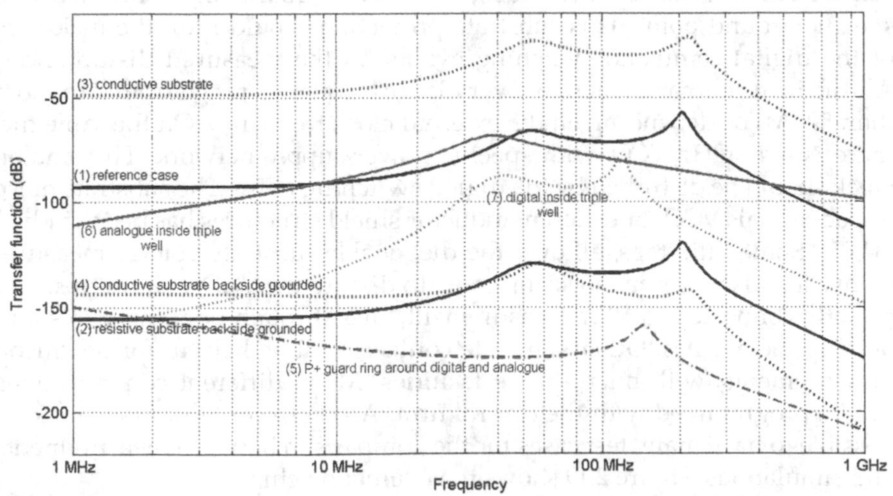

FIGURE 2.18
Comparative substrate isolation study on a virtual test case study: transfer functions for the different configurations.

provided by the P+ guard-ring technique: the guard-rings are biased with perfect Kelvin points in this case and dissipate as best as possible the substrate perturbations. Moreover, a triple-well isolation may be very efficient in isolating a small area block, but considering now a large area block, the parasitic capacitance between N-well and substrate will be larger and the cutting frequency smaller.

2.6 Application to a Mixed-Signal IC

2.6.1 Test-Chip Presentation

To validate the proposed methodology and to investigate the substrate isolation technique, a test chip was realized in a 0.35 μm BiCMOS SiGe process with five metal/two poly layers and a high-resistive substrate (~15 Ω cm). The circuit supply voltage is 3.3 V. This test chip should allow us to characterize substrate noise generated by the digital part of this kind of small mixed-signal IC. The package used is a QFN 5 × 5 mm^2 with 28 pins. It has been chosen for its very short bonding wires and its backside connection. This kind of package is currently proposed for RF and low-noise applications. On the die, four inverter networks are placed to generate noise in the resistive substrate. They can be driven, with many configurations, by a programmable digital command. We can drive different numbers of inverters according to their drive or their location on the layout. The inverter network control unit is isolated in the substrate by a P+ guard-ring connected to a specific ground point. This substrate protection should limit the incidence of the digital command switching events on the measured disturbances. All the inverters can switch in a very large frequency range from 0 to more than 500 MHz, depending on the external clock frequency. On the same die, there is a 4.5 GHz VCO, with specific power-supply network. This analog block should be disturbed by the digital switching noise. The substrate noise impact on this VCO for different inductor shield structures has been studied [34,35]. Some bulk taps, all over the die, enable substrate voltage measurements and also external noise injection, to disturb the VCO. Four of them are directly connected to 50 Ω measurement lines through bonding wires and on-chip metal lines. Other substrate contacts will be helpful for on-silicon measurements, with micro-probe facilities. Many different configurations can be programmed with the control unit. All these configurations make it possible to have many test cases for the comparison between measurements and simulations. Figure 2.19 shows the evaluation chip.

A socket is also placed on the evaluation board (Figure 2.20). This socket allows the quick change of the device under test but also introduces other parasitic elements in supply and measurement lines.

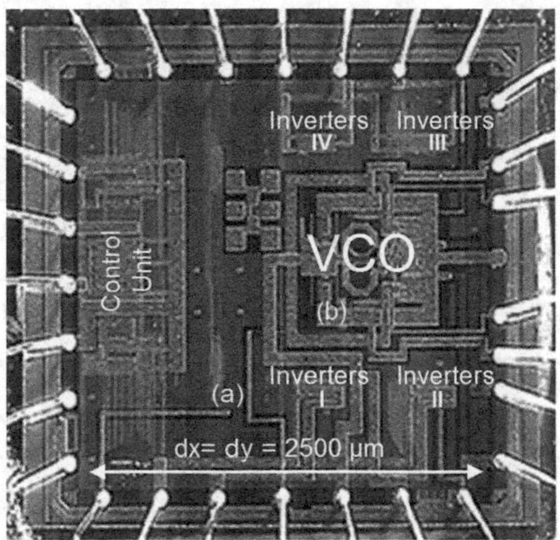

FIGURE 2.19
Microphotograph of evaluation chip.

FIGURE 2.20
Evaluation printed circuit board.

The measurement method is quite simple compared to other papers [33,36–47] and surely more sensitive to other coupling phenomena like capacitive or electromagnetic coupling. With a very efficient measurement lines modeling and a good choice of substrate contacts to probe, we can conclude from the waveforms measured the realistic image of the parasitic substrate voltage.

The first version (normal version) was characterized and allowed us to validate the modeling methodology [19]. The different parts that have to be modeled to create the extended-ICEM model are described in Figure 2.21.

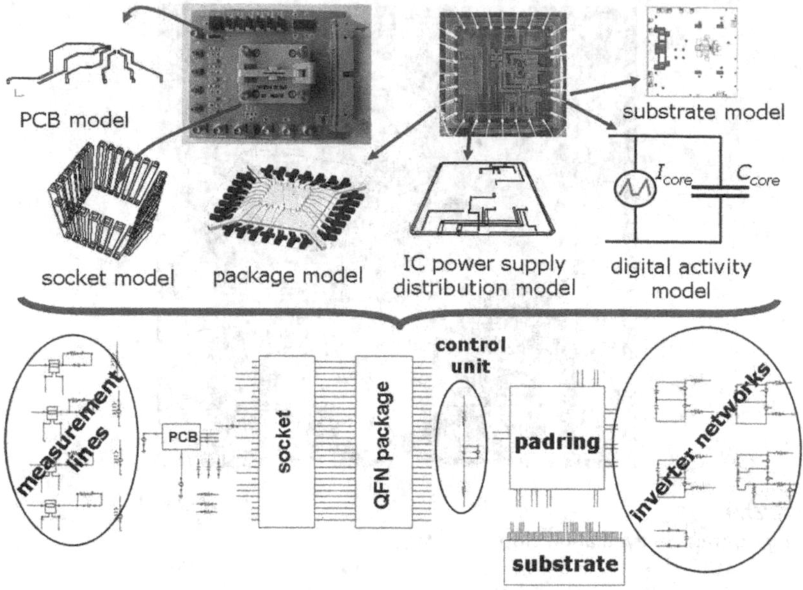

FIGURE 2.21
Building the complete model (based on ICEM approach with substrate extension).

2.6.2 Substrate Noise Reduction: Low-Noise Version

The challenge is now to develop a low-noise version. The first effort concerns the noise generator itself. As already mentioned, dynamic signals involve dynamic current consumption and so high current spikes. By reducing the average activity of the digital device, we can decrease the global noise generation [33,37–39]. In our case, we have chosen to decrease the dynamic activity of the digital part by enlarging the digital switching activity timing window. This window is called the skew. It is the time between the first logical gate commutation and the last one during a digital clock period. Increasing this skew value allows the digital circuit to absorb its necessary energy within a longer period of time. Thus, the global consumption current is less dynamic and involves less interference in the power-supply network. To apply this method to our circuit, we have changed the commutation delay of each inverter network, in such a way that they do not switch at the same time.

Secondly, we investigate the use of isolation techniques such as P guard-rings. In both cases, when the guard ring is around the digital or the analog part, the simulated substrate parasitic levels are as high as the non-isolation configuration. The reason is that to be effective, the P+ guard ring has to be connected to the ground as cleanly as possible. In our circuit, the ground line parasitic elements do not allow to correctly evacuate the parasitic substrate currents.

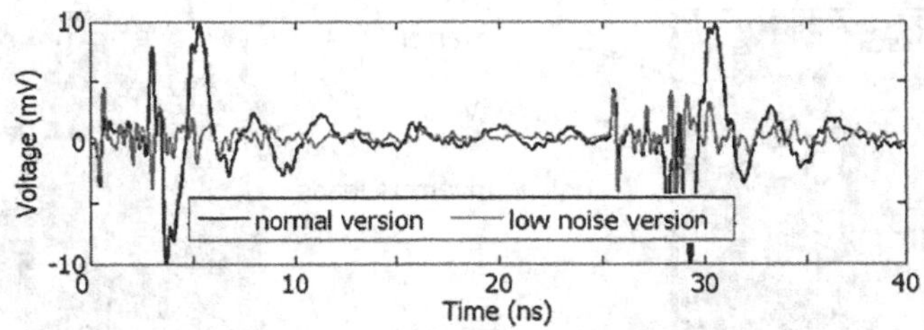

FIGURE 2.22
Measurements of parasitic substrate noise on triple-well isolated substrate contact.

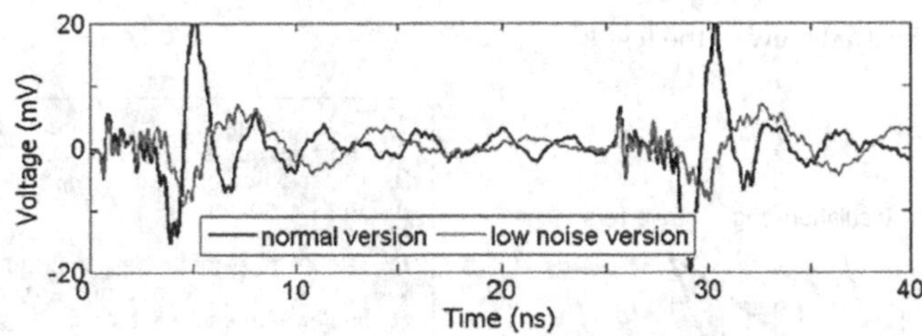

FIGURE 2.23
Measurements of parasitic substrate noise on common-substrate contact.

Finally, the passive distribution network is optimized. The on-silicon capacitance has been increased, to filter the on-silicon natural power ringing. A power-supply resistance has also been added, in order to absorb all the oscillations.

Therefore, a second version of the test-circuit has been implemented [19]. An embedded capacitance of 1 nF and a serial resistance on power supply of 10 Ω have been added to the normal test-chip version. Additionally, a substrate measurement tap has also been enclosed in a triple-well. The new version measurements are given on Figures 2.22 and 2.23, for two different substrate measurement taps: the first one is closed inside N well, the second one is in the same substrate as the perturbing digital blocks.

The RMS values of the perturbing signals are the following: 3.8 and 7.1 mV for the normal version of the circuit; 1.4 and 3.2 mV for the low-noise version. As expected, the perturbation frequencies are reduced due to the new value of the on-silicon capacitance. The *RLC* oscillations are also dampened by the power-supply serial resistance.

2.6.3 Voltage-Controlled Oscillator Spectrum in Normal and Low-Noise Version

The integrated in the test chip (see Figures 2.24–2.27) is based on a cross-coupled differential pair of hetero-junction bipolar transistors [48–54] with a capacitive bridge feedback. The tuning voltage can vary from 0 to 3 V to

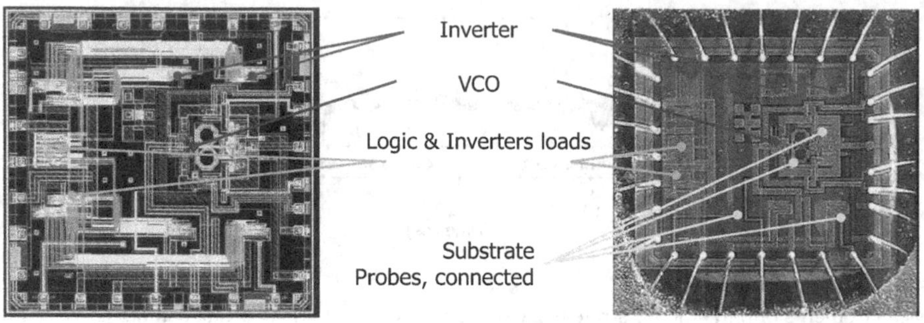

FIGURE 2.24
CAD view and actual view of the test IC.

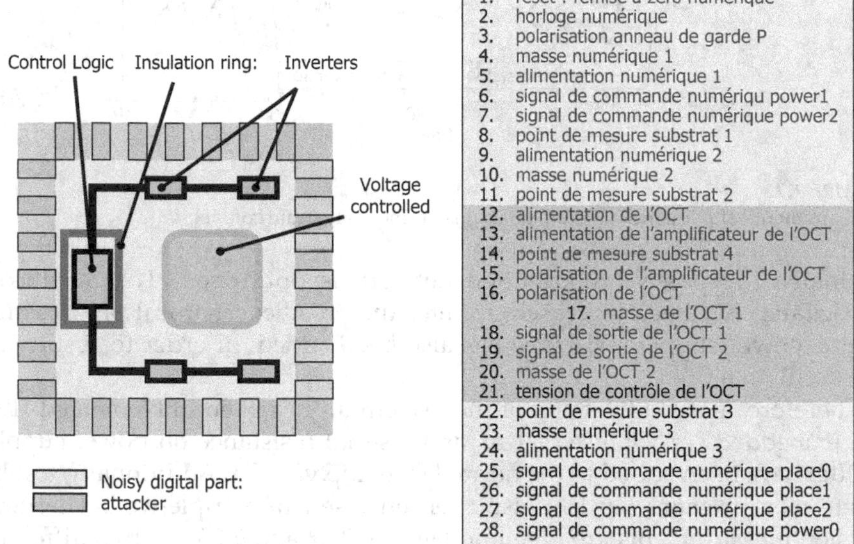

FIGURE 2.25
Localization of attacker and target parts in the VCO.

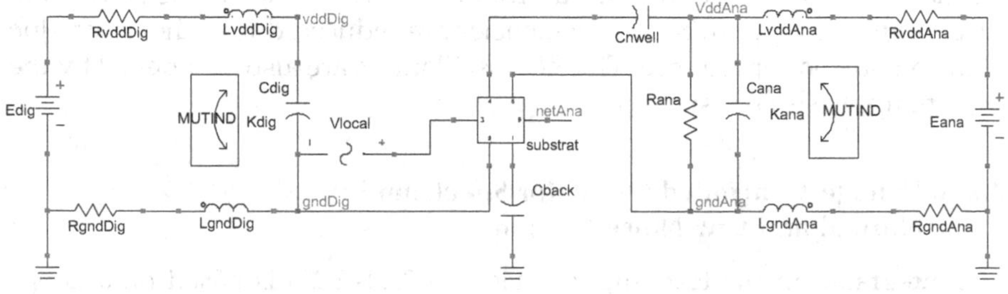

FIGURE 2.26
Simulation electrical schematic of the local substrate noise local on the basic configuration.

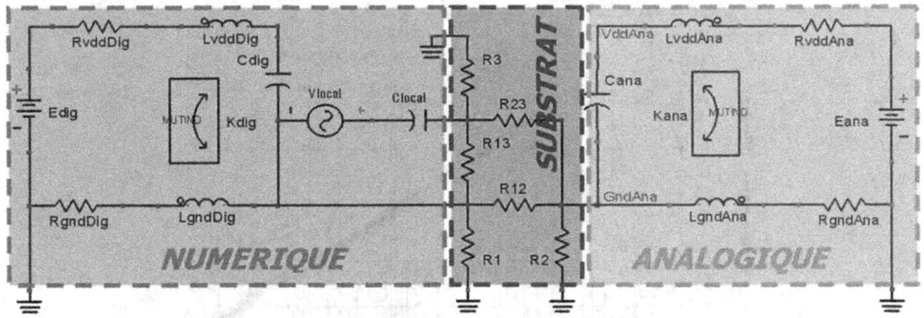

FIGURE 2.27
Total electric diagram of simulation of the disturbances substrate generated on the mass of the analogical part of a mixed circuit by the local phenomena.

obtain a carrier frequency in the range of 4.30–4.55 GHz. The output signals pass through buffers in a common collector circuit configuration. The maximum carrier power with a 50 Ω load is close to 0 dBm (measured on the SMA output connector). The phase noise measured at 100 kHz from the carrier varies from –90 to –100 dBc/Hz (depending on-chip dispersion and biasing). Two substrate taps have been placed inside the VCO core in order to inject a parasitic signal directly into the VCO substrate, with an external source.

The VCO spurious side-bands (Figure 2.28) caused by harmonic substrate noise disturbances have been analyzed in [3,19].

Close to the carrier, the phase noise can deteriorate when the digital part is active. As an illustration, Figure 2.29 presents the phase noise measurement for the normal test-chip version superposed with low-noise version with the clock frequency at 10 kHz. Due to frequency conversion, the clock frequency and its odd harmonics are observed in the VCO phase noise. Spur attenuation between 5 and 6 dB is measured for the low-noise-version, confirming the digital noise reduction.

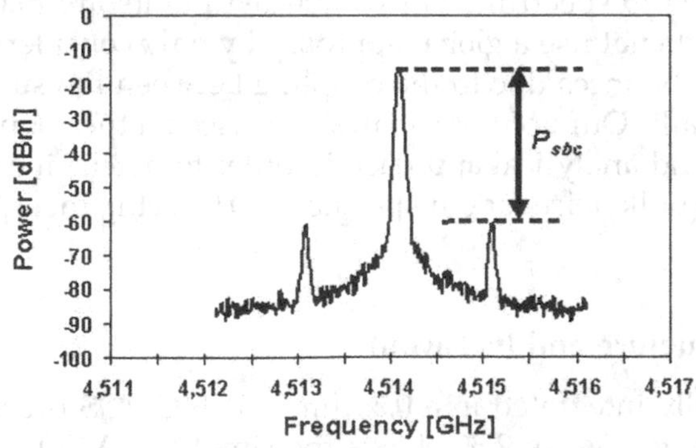

FIGURE 2.28
Measured P_{sbc} for a small sine-wave applied to the substrate contact near the inductor (50-mV peak at 1 MHz).

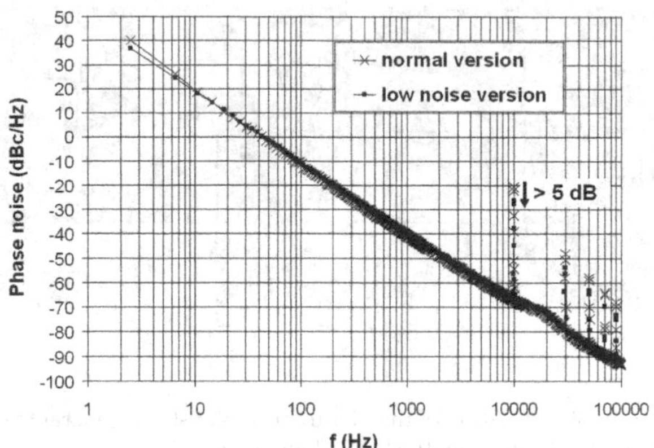

FIGURE 2.29
VCO phase noise measurement with $f_{clock} = 10\,kHz$ (normal and low-noise version).

2.7 Impact of Low-Frequency Substrate Perturbations on an RF VCO Spectrum

We investigate VCO sensitivity to low-frequency substrate noise (low frequency compared to the VCO carrier frequency). We analyze the way the substrate noise (internal digital noise or injected noise into the chip) is converted close to the oscillator carrier (some spurious side-bands appearing) and may impact the oscillator spectrum quality. Simple calculations and measurements without the complete phase noise formalism are possible as we quantify the coupling mechanism by evaluating side-bands power relative to the carrier magnitude (in dBc). Compared to a previous work [40–42], this one aims to locally analyze coupling mechanisms between substrate noise and the VCO spectrum as two injected points are placed inside the VCO core. We do not use a global approach by only considering power and ground supply bounces due to the coupling between the substrate and the supply metal rails. Our accurate analysis focuses on the coupling with both experimental and analytical approach in order to determine which devices are sensitive to bulk noise for our specific VCO test chip including integrated injection taps.

2.7.1 VCO Structure and Its Layout

The VCO is fully integrated in a 0.35 μm SiGe BiCMOS process with high-resistive substrate. Figure 2.30 shows the simplified VCO schematic view. The main part of the VCO is the LC-tank. The negative resistance is obtained by a cross-coupled differential pair of hetero-junction bipolar transistors (HBT) biased by I_{bias} current.

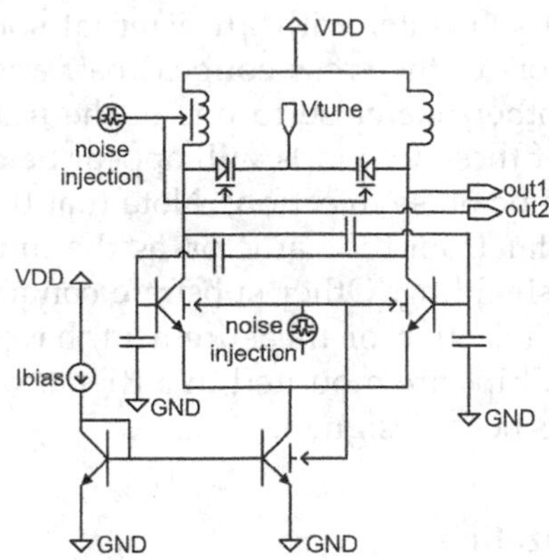

FIGURE 2.30
Simplified VCO core with injection contacts.

To avoid frequency variations due to supply voltage noise, many on-chip decoupling capacitors are placed in the VCO (see layout in Figure 2.31). For the same purpose, substrate contacts and guard-rings surrounding the devices presenting a large surface in contact with the bulk have been added.

The circuit supply voltage is 3.3 V. The tuning voltage can vary from 0 to 4 V to obtain an oscillation frequency in the range [4.25–4.60 GHz]. The output signals pass through buffers in a common collector circuit configuration. The maximum carrier power with a 50 Ω load is closed to 0 dBm. The phase noise measured at 100 kHz from carrier is varying from –90 up to –100 dBc/Hz (depending on-chip dispersion and biasing).

Test-chip microphotograph is shown in Figure 2.31. The chip contains a VCO, digital blocks, and many substrate taps. Especially two substrate taps have been placed inside the VCO core in order to inject a parasitic signal

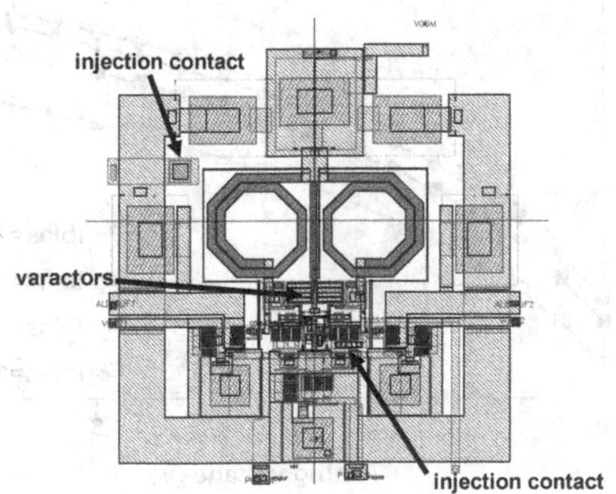

FIGURE 2.31
VCO layout.

directly into the VCO substrate, with an external source: one tap in the vicinity of the transistors of the cross-coupled pair and the current mirror (I_{bias} source), and the other one close to one of the inductors (Figures 2.30 and 2.31). The utility of these two taps will appear below (side-bands measurements with harmonic noise injection). Note that the substrate tap close to the inductor is not far from the varactor, as the inductor does not have any substrate pattern shielding. Other substrate contacts (outside the VCO core); for perturbation injection or measurement, have different areas and locations on the chip. Chips are mounted in a RF package (VFQFPN) and a dedicated RF board has been designed.

2.7.2 VCO Characterization

A. *Static sensitivity functions: VCO carrier frequency variations, function of bias current, and tuning voltage*

VCO carrier frequency depends on inductance and varactor choice, but also on bias current or tuning voltage. Figure 2.32 shows the measured VCO frequency variations as a function of the tuning voltage and the bias current. These variations of the VCO carrier frequency f_c, function of the bias current or tuning voltage, can be described by some VCO sensitivity functions, respectively, called K_{bias} (MHz/mA) and K_{tune} (VCO gain in MHz/V), that is, $K_{bias} = \partial f_c / \partial I_{bias}$ and $K_{tune} = \partial f_c / \partial V_{tune}$. We extract the VCO sensitivity functions by using derivative of the frequency-dependent curve relative to the tuning voltage or the bias current; Figures 2.33 and 2.34 represent, respectively, the sensitivity functions of bias current (K_{bias}) and tuning voltage (K_{tune}).

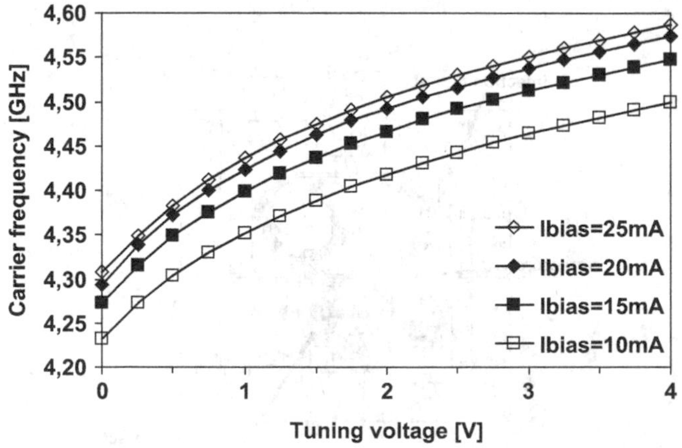

FIGURE 2.32
VCO carrier frequency as a function of the tuning voltage, for different bias currents.

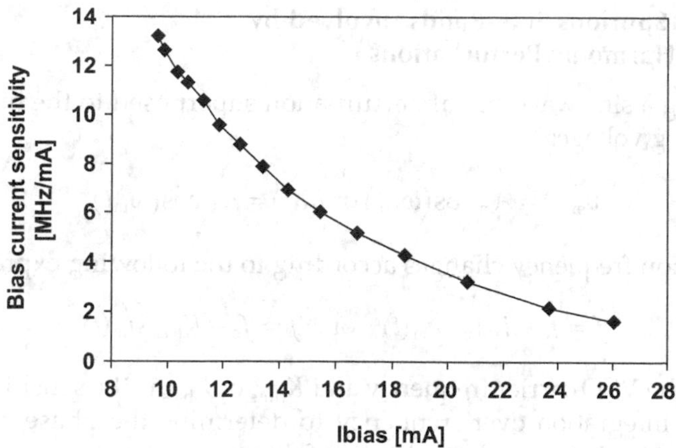

FIGURE 2.33
Bias current sensitivity function (K_{bias}), with $V_{tune} = 2\,\text{V}$.

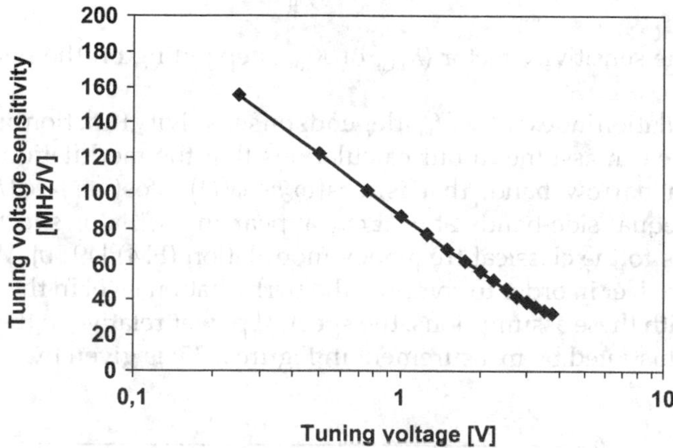

FIGURE 2.34
Tuning voltage sensitivity function (K_{tune}), with $I_{bias} = 17\,\text{mA}$.

We observe that the variations of the bias current results in a small change of the oscillator frequency (Figure 2.33). The voltage disturbance applied to varactors generates a sensitivity level much more important than the bias current sensitivity (Figure 2.34). Roughly, the sensitivity function magnitude for bias current K_{bias} is ten times lower compared to tuning voltage sensitivity K_{tune}. This method lets us establish the VCO sensitivities as functions of circuit bias conditions by measurements or simulations. In the next section, we analyze the impact of small signal perturbation superposed to bias current or tuning voltage. The VCO sensitivity to VDD-supply voltage (K_{VDD}) can also be measured or simulated; then K_{GND} (VCO sensitivity to on-chip ground voltage) can be extracted, according to the relation: $K_{GND} = -(K_{VCO} + K_{VDD})$ [19].

2.7.3 VCO Spurious Side-Bands Involved by Bias Harmonic Perturbations

Considering a sine wave signal perturbation superposed to the bias current or the tuning voltage:

$$v_{\mathrm{m}}(t) = A_{\mathrm{m}}\cos(\omega_{\mathrm{m}}t) \text{ or } i_{\mathrm{m}}(t) = A_{\mathrm{m}}\cos(\omega_{\mathrm{m}}t) \tag{2.1}$$

the oscillation frequency changes according to the following expression:

$$f = f_{\mathrm{c}} + K_{\mathrm{tune}} \cdot v_{\mathrm{m}}(t) \quad \text{or} \quad f = f_{\mathrm{c}} + K_{\mathrm{bias}} \cdot i_{\mathrm{m}}(t) \tag{2.2}$$

where f_{c} is the VCO carrier frequency and K_{tune} or K_{bias} is the sensitivity factor.

After an integration over frequency to determine the phase, the output signal of the oscillator can be written as follows:

$$s(t) = S \cdot \cos\left(\omega_{\mathrm{c}}t + \frac{A_{\mathrm{m}}K}{f_{\mathrm{m}}}\sin(\omega_{\mathrm{m}}t)\right) \tag{2.3}$$

with K is the sensitivity factor (K_{bias} or K_{tune} depending on the perturbation location).

The modulation index ($A_{\mathrm{m}}K/f_{\mathrm{m}}$) depends on sensitivity function but remains small; so we can assume in our calculations that the modulation is always placed in a narrow band, that is, $\cos(\omega_{\mathrm{c}}t + \varphi(t)) \approx \cos(\omega_{\mathrm{c}}t) - \varphi(t)\sin(\omega_{\mathrm{c}}t)$. Therefore, equal side-bands at $\omega_{\mathrm{c}} \pm \omega_{\mathrm{m}}$ appear in oscillator spectrum, that corresponds to the classical frequency modulation (FM) [49,50]. We refer to the carrier power in order to compute the perturbation level in the frequency domain. With these assumptions, the spectral power relative to the carrier in dBc unit (illustrated by measurement in Figure 2.35) is given by

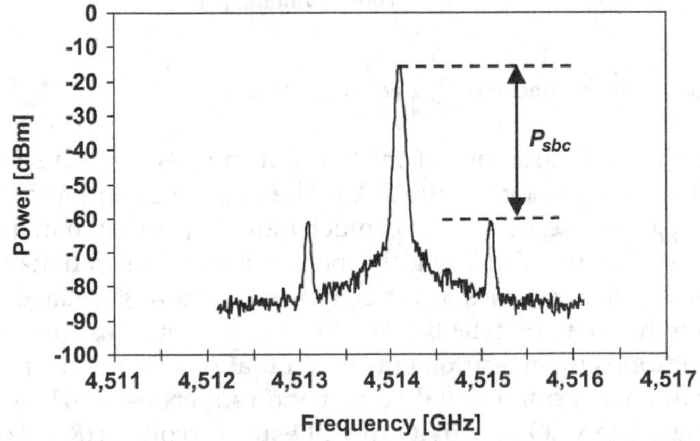

FIGURE 2.35
Measured P_{sbc} for a small sine-wave applied to the contact near inductors (50 mV peak at 1 MHz).

$$P_{sbc} = 10 \cdot \log\left[\left(\frac{K \cdot A_m}{2f_m}\right)^2\right] \tag{2.4}$$

A perturbation superposed to the bias current or tuning voltage generates spurious side-bands that have the power magnitude directly proportional to *K*-sensitivity. In the same time, the *K*-sensitivity is function of the bias current and of the tuning voltage. Thus, the reduction of the noise power below carrier may be done by a proper choice of the biasing or tuning.

2.7.4 VCO Spurious Side-Bands Involved by Substrate Harmonic Perturbations

Considering now a substrate perturbation, the difficulties appear numerous. An experimental approach demands to design many dedicated test-chips to analyze coupling phenomena. Simulations are also possible, with at first, the quantification and location of the perturbation signal (noise sources). Different studies focus on modeling injected noise by CMOS digital part in ICs [2,19]. Once equivalent noise sources are determined, these late ones are associated in a global scheme describing the board, the IC socket and package, the bondings, the pad ring, the power-supply rails and the coupling to substrate (resistive and capacitive) [39,40]. The next step concerns the propagation modeling. Designers can use substrate extractor (post-layout extraction) [20,44–47] to obtain a substrate *RC*-network compatible with some SPICE-like simulator. Such very large extracted netlists are not easy to handle since the simulation duration drastically increases (with needed memory space). Moreover such extractions are always associated with an *RC* reduction step limiting the frequency validity range. The last point is the study of the sensitive bloc (e.g., the VCO in our case). Accurate simulations are possible by means of precise models for passive and active devices (including substrate effect, of course) coupling with advanced simulation methods [2].

Experimentally, when sine-wave voltage is applied to substrate taps (inside the VCO core), we measure VCO side-bands (spurs) closed to the carrier frequency f_c (Figure 2.35). As explained in the previous section, for a harmonic noise at f_m frequency, we observe side-bands at $f_c \pm f_m$ (the FM conversion mechanism is verified as spur magnitude is proportional to noise frequency). The VCO sensitivity, due to the injected perturbation (level, location, frequency, etc.), is quantified by P_{sbc} values (Figures 2.35 and 2.36). Figure 2.36 gives an example of P_{sbc} measurement for a small sine-wave applied to the contact near inductors (50-mV peak at 150 kHz). Figure 2.36 presents the measured P_{sbc} values for a sine-wave perturbation injected closed to the inductor/varactors and the bipolar transistors of the differential pair (150-mV peak at 150 kHz). From this figure, we observe firstly that side-band power values (P_{sbc}) are larger when the injection is located closed to bipolar transistors of the cross-coupled differential structure and the current mirror; on the other

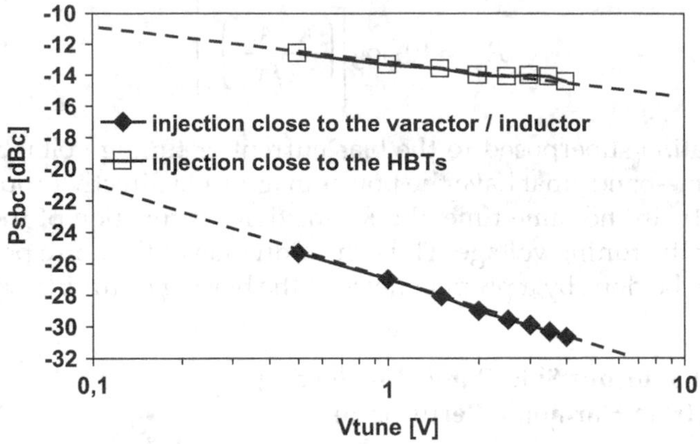

FIGURE 2.36
Measured P_{sbc} values function of the tuning voltage (150mV peak at 150kHz perturbation, injection close to inductors or bipolar transistors).

hand, the decrease of P_{sbc} values with tuning voltage is more significant when injection occurs near inductors (Figure 2.35). In the rest of the chapter, we try to analyze these two key observations: a larger impact when injection is located close to the bipolar transistors and an impact depending on VCO tuning voltage when injection is located close the inductors (and varactors).

Using a global approach, we can express a VCO carrier frequency change as follows:

$$\Delta f_c = K_{bias}\Delta I_{bias} + K_{tune}\Delta V_{tune} + K_{VDD}\Delta V_{VDD} + K_{GND}\Delta V_{GND} + K_{sub}\Delta V_{sub} \quad (2.5)$$

where K_{sub} and ΔV_{sub} take into account the VCO carrier frequency sensitivity to substrate voltage.

When substrate noise is present, we can consider firstly, a propagation from substrate to on-chip ground through ground metal rail (K_{GND}, ΔV_{GND}) and secondly, a direct substrate path (K_{sub}, ΔV_{sub}). The others terms in (5) can be neglected, due to the weak capacitive coupling at low substrate noise frequencies. In the next section, we investigate what are the sensitive devices to substrate noise, using simulated impulse response to some substrate perturbation.

2.7.5 Analysis of Substrate Coupling Mechanisms by Means of the ISF Approach

2.7.5.1 ISF Principle

In this section, we use the linear time-variant model described in [41] to analyze the VCO sensitivity to substrate perturbations. The method is based on the Impulse Sensitivity Function (ISF) calculation which represents the excess phase after applying some perturbation impulse to an oscillator circuit. Varying the perturbation impulse event time (τ) during an oscillator period, we can access to the ISF function $\Gamma(\tau)$. Note that we do not consider the magnitude shift as it disappears with time; on the other hand the phase shift is preserved. The perturbation impulse can be a current spike into a capacitive node (injected charge), or a voltage spike onto an inductive node. The ISF function $\Gamma(\tau)$ is dimensionless and has the same period as the oscillator one (or one can consider that $\Gamma(\omega_c\tau)$ has a 2π-period) [41]. For the sake of clarity, we will consider the pseudo-ISF function $\Gamma^{\varphi}(\tau)$ in the following lines of this section. $\Gamma^{\varphi}(\tau)$ has the dimension of radian/As or radian/Vs depending on the type of perturbation (current or voltage). Finally, we can write the excess phase for an impulse response as follows:

$$h_{\phi}(t,\tau) = \Gamma^{\varphi}(\omega_c\tau) \cdot u(t-\tau) \tag{2.6}$$

where $u(t)$ is the unit step.

In other words, $\Gamma^{\varphi}(\tau)$ function is a direct representation of the excess phase (phase shift), normalized by injected charge (for current impulse). Due to its periodicity, $\Gamma^{\varphi}(\tau)$ function can be extended in a Fourier series as follows:

$$\Gamma^{\varphi}(\omega_c\tau) = \frac{c_0}{2} + \sum_{n=1}^{\infty} c_n \cdot \cos(n\omega_c\tau + \theta_n) \tag{2.7}$$

Considering, now, the harmonic perturbation (current or voltage) defined by its magnitude A and its angular frequency ω_m (frequency: f_m), that is:

$$p(t) = A \cdot \cos(\omega_m t) \cdot u(t - t_0) \tag{2.8}$$

Thanks to the linear time-variant system, the phase shift is straightly obtained [41] according to the following expression:

$$\varphi(t) = \int_{-\infty}^{t} \Gamma^{\varphi}(\omega_c\tau) \cdot p(\tau) \cdot d\tau = \int_{t_0}^{t} \Gamma^{\varphi}(\omega_c\tau) \cdot A \cdot \cos(\omega_m\tau) \cdot d\tau \tag{2.9}$$

Then, the phase shift can be written as follows:

$$\varphi(t) = A \cdot \frac{c_0 \sin(\omega_m t)}{2\omega_m} + A \cdot \sum_{1}^{\infty} \frac{c_n \sin\left[(n\omega_c \pm \omega_m)t + \theta_n\right]}{2(n\omega_c \pm \omega_m)} + \varphi_0(t_0) \tag{2.10}$$

In (10), only the term calculated with $\omega_m = n\omega_c + \Delta\omega$ (with $\Delta\omega \ll \omega_c$ and $n = 0$, 1, 2...) is to be taken into account in the final expression of the oscillator output $s(t) = S \cdot \cos(\omega_c t + \varphi(t))$, while all the other terms are negligible. Finally, only low-frequency noise (n = 0) and noise disturbance around harmonics (n = 1, 2...) impact the phase. The c_n coefficient of the ISF function is used to calculate the phase shift introduced by the noise at pulsation $n\omega_c + \Delta\omega$. Then, the oscillator output can be written as follows:

$$s(t) \approx S \cdot \left[\cos(\omega_c t) - \varphi(t)\sin(\omega_c t)\right] = S \cdot \left[\cos(\omega_c t) - \frac{c_n A}{2\Delta\omega}\sin(\Delta\omega t)\sin(\omega_c t)\right] \tag{2.11}$$

According a narrow side-band condition (already described in Section III.B), a noise at $n\omega_c + \Delta\omega$ results in a pair of equal side-band at $\omega_c \pm \Delta\omega$ with a side-band power relative to the carrier, equals to:

$$P_{sbc}(\Delta\omega) = 10 \cdot \log\left[\left(\frac{c_n A}{4\Delta\omega}\right)^2\right] \tag{2.12}$$

Note that we obtain a unique P_{sbc} (in dB_c) expression (2.12) for any harmonic, by using $c_0/2$ term in (2.7), instead of c_0.

2.7.5.1.1 *ISF Applied to an Injected Perturbation on Varactor Substrate*

The ISF function $\Gamma^{\varphi}(\tau)$ is calculated with a current perturbation directly applied to the substrate pins of the two varactors (Figure 2.37). As an illustration, we obtain the ISF curve (Figure 2.37) by simulating the phase shift (excess phase) occurring when the injected charge timing event τ is sweeping over an oscillator period. The simulations are performed with a SPICE-like simulator (time-domain simulations) with a sufficient duration to be sure to obtain a stable phase shift [41]. The current pulse (1 mA magnitude and 10 ps duration) is simultaneously applied to the substrate pins of the

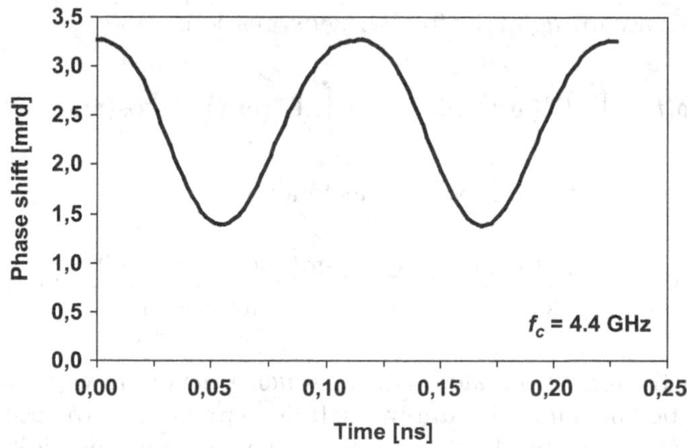

FIGURE 2.37
ISF function (pulse 1 mA, 10 ps) for injection in the substrate pins of varactors (V_{tune} = 3 V).

two varactors. We have also verified the linearity, not illustrated here (excess phase proportional to injected charge). We note that the ISF varies with twice the oscillator frequency (already reported in [20,41]).

To validate the ISF approach, we compare the P_{sbc} values calculated with (3.43) (using the c_n coefficients) with the values obtained directly by means of time-domain simulation coupled with discrete Fourier transform (DFT) post-processing analysis. These comparisons are shown in Figure 2.38 for a 200 µA harmonic current noise at $nf_c + \Delta f$ with n: harmonic and $\Delta f = 50\,\text{MHz}$.

The simulated waveform for P_{sbc} is shown in Figure 2.38 for a current harmonic injected at 50 MHz with 200-µA magnitude. As already mentioned, the frequency domain curve is obtained with a DFT post-processing analysis, with a very long transient time analysis to be sure to obtain the periodic steady state. The curve is normalized using arbitrary 0 dBm reference power magnitude for the VCO carrier frequency ($f_c \approx 4.4\,\text{GHz}$); then we directly observe that P_{sbc} equals to –25 dBc for $n = 0$ (Figure 2.39) for the two side-bands at $f_0 + 50\,\text{MHz}$ and $f_0 - 50\,\text{MHz}$. This value can be compared to the one calculated by ISF approach. This is derived by calculating the c_n Fourier coefficients of the ISF function (Figure 2.40) with (12). We extract $c_0 = 0.0048$ (we remind that the pulse characteristics for this ISF calculation were 1 mA, 10 ps). Using (12), we obtain for the current harmonic 200 µA at 50 MHz, a P_{sbc} equals to –22 dBc.

As this first use of the ISF approach for noise injection into the varactor substrate pin seems quite satisfactory, we analyze the device sensitivity to substrate noise in the next section by calculating excess phase function of bias conditions and the injected noise location.

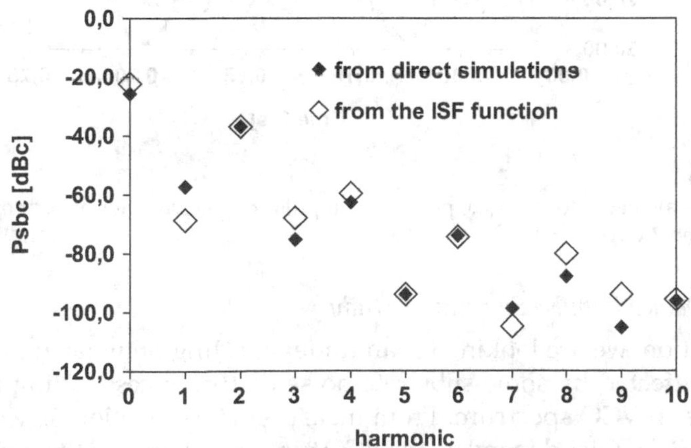

FIGURE 2.38
P_{sbc} comparisons between direct transient simulations and the ISF approach (200 µA harmonic current noise at $nf_c + \Delta f$ with n: harmonic and $\Delta f = 50\,\text{MHz}$).

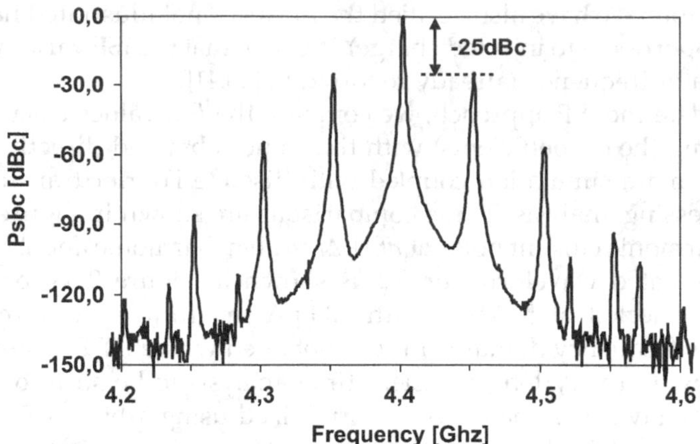

FIGURE 2.39
P_{sbc} determination for harmonic $n = 0$ (current harmonic 200 μA at 50 MHz): transient simulation and post-processing DFT analysis.

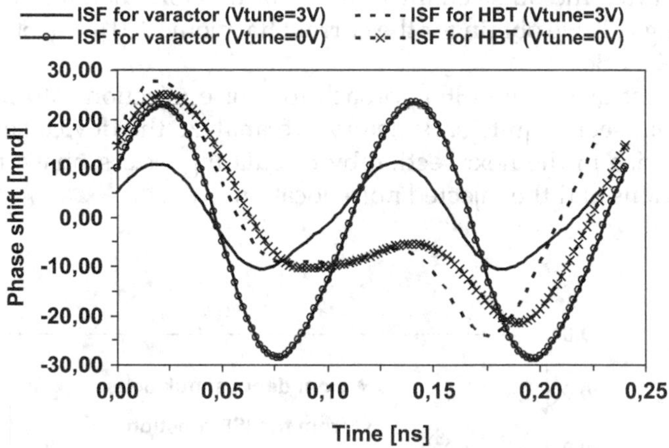

FIGURE 2.40
ISF function calculated for a 1 V/20 ps voltage impulse on one varactor bulk or one HBT bulk ($V_{time} = 0$ V and 3 V).

2.7.5.1.2 *Device Impulse Sensitivity Study*

In this section, we are looking for an understanding on what are the devices the most affected by some substrate noise disturbances leading to observe side-bands in VCO spectrum. From measurements (Section 3), we note that measured P_{sbc} values depend on the VCO tuning voltage and the injected noise location inside the VCO (location close to one of the inductors or in the vicinity of the HBTs of the cross-coupled pair and the current mirror). Substrate noise induces GND power-supply bounces (V_{tune} and VDD power-supply bounces are negligible due to the weak capacitance coupling). In our specific test chip, considering a high-resistive substrate (noise attenuation with distance: the substrate cannot be considered as a single node), numerous on-chip decoupling capacitors, injection nodes inside VCO; on one hand, we can consider a propagation from injection substrate tap to device substrate through the ground metal rails and, on the other hand, a direct path between injected nodes and passive or active device bulk (varactor, inductor, HBTs, etc.). This is the key point we want to analyze by means of the ISF approach. Then we simulate phase shift evolution by plotting excess phase variations, function of V_{tune}, for a voltage pulse or a harmonic perturbation applied directly to the different device bulk pins. A lot of time-domain simulations have been running and analyzed. Hereafter we summarize the main points:

- Due to the differential structure, pulse injection in a single bulk node is most significant (compared to simultaneous injection in symmetrical device bulks).
- Mainly varactors and HBTs are concerned by substrate voltage disturbances Inductors, MIM capacitors, and polysilicon resistors seem not to be influenced by low-frequency substrate noise.

In this section, we do not use a substrate extraction tool and we simulate the perturbation directly applied to the device bulk pin. The very reason is, that using a complete substrate RC extracted netlist, we obtain a false carrier frequency due to the limited frequency validity range after extraction, that is, substrate reduced netlist is not further valid up to a few GHz. Figure 2.40 presents the ISF curves for the highest sensitive devices; ISF are calculated when a voltage impulse (1 V during 20 ps) is applied to one varactor bulk or one HBT bulk (both calculated for $V_{tune} = 0\,V$ and 3 V). Finally, on Figure 2.41, we show simulated P_{sbc} values function of tuning voltage for a harmonic voltage noise at 50 MHz (10 mV peak magnitude), also applied to the same varactor bulk and HBT bulk. These two figures are particularly interesting as, first of all, they are correlated as it will be explained here after and secondly, they agree with measurements. The complete study of the ISF function is not developed here (Fourier series), but comparing the ISF curves of Figure 2.40, we observe that the ISF curve for the varactor has twice the carrier frequency and is symmetrical for $V_{tune} = 3\,V$ (small $c_0/2$ mean value). On the other hand, the ISF curve for HBT has the same period as the oscillator one but it presents some dissymmetry leading to a significant $c_0/2$ term in (7). In other words, for a 3 V tuning voltage, the impact in the low-frequency noise range is more important for the HBT than the varactor (12). This conclusion is drastically different for high-frequency noise, around carrier frequency or twice carrier frequency as the Fourier coefficients differ (not discussed here). We also observe that the ISF curve for the varactor, at $V_{tune} = 0\,V$, is higher and dissymmetrical, leading to an increase of the $c_0/2$ mean value. This is correlated to the results of II, as shown in Figures 2.36 and 2.41, where P_{sbc}

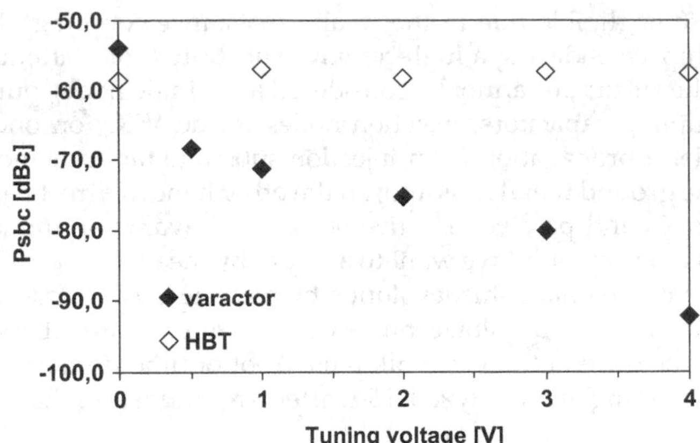

FIGURE 2.41
Determined P_{sbc} values function of the tuning voltage by means of the ISF approach (voltage harmonic noise 10 mV peak @ 50 MHz injected on one varactor substrate pin or one HBT substrate pin).

values are higher for a 0 V tuning voltage. Finally, the calculated P_{sbc} values (Figure 2.41) agree well with the measured ones (Figure 2.36); P_{sbc} values for injection close to HBT do not depend on the tuning voltage. For injection close to the inductor, the coupling behavior is driven by the varactor sensitivity and we observe P_{sbc} decreasing with the tuning voltage. P_{sbc} magnitudes are higher for injections close to the HBTs than for injection in the vicinity of the inductor, because the attenuation through substrate decreases the noise magnitude at the varactor bulk. In a future work, we will study the same structure with a shielded inductor (substrate pattern) in order to modify the substrate direct path between injection location and varactor.

2.8 Conclusion

In this chapter, resumed are the basics of the substrate noise generation and propagation in mixed-signal ICs. We also present the standard ICEM model and its extension allowing substrate noise simulations. This model considers only the noise generated in the substrate by the power-supply and ground fluctuations of the digital part. Although this approach seems very simple, it gives good results regarding test-chip measurements. To reduce substrate noise, the designers can control some parameters concerning, first of all, the noise generator (digital block). The skew and the transition time can be optimized in order to reduce the average current and therefore power-supply ringing. The passive power-supply distribution network, through its parasitic elements *RLC*, has also a key role. Adding decoupling capacitance (PCB and/or on-chip), resistance (if possible) may reduce oscillation frequency and magnitude. Finally, some substrate isolation techniques can be implemented (beginning with the substrate resistivity choice and the die backside connectivity). Classical guard-rings and triple-wells are always solutions but their efficiency depends on the biasing quality. Furthermore, triple-well will lose its isolation property when its area increases.

General rules cannot be claimed as each design is a specific case and the solutions have to be adapted. Our contribution is pragmatic as it proposes a simple investigation methodology to a complex problem. Nevertheless, some work remains to be continued, for example concerning the way to obtain the transient consumption current of an entire large digital part or an efficient method to simplify a layout and to extract an accurate substrate *RC* network. Moreover, the signal integrity issue in SoC or Systems-in-Package (SiP) does not only concern conducted noise but also radiated propagation with electromagnetic effects: the ICEM standard model can also respond to this need.

The impact of low-frequency substrate noise on a 5-GHz voltage-controlled oscillator spectrum is a key example. First of all, we present the test-chip structure with a fully integrated VCO and two substrate taps inside the VCO core for noise injection or measurement (one close the inductor/varactor and the other one in the vicinity of the bipolar transistors of the cross-coupled pair and the current mirror). The VCO characterization is performed by measuring static sensitivity curves that represent the frequency variations, function of the tuning voltage or the bias current.

The next step is the study of the sensitive devices using the ISF. Many simulations have been carried out leading to conclude that mainly the varactors and bipolar transistors are affected by substrate noise. For high-frequency substrate noise, the behavior can be analyzed through the ISF Fourier series. In a realistic IC, noise source locations, layout implementation (power-supply rails, shielding, guard ring, etc.) and substrate attenuation have to be taken into account to quantify the noise magnitude on each device substrate.

3

Efficient and Simple Compact Modeling of Interconnects

3.1 Introduction: Overview of the 3D Interconnect Modeling Approach

A significant physical design challenge in both high-performance three-dimensional (3D) integrated circuits (IC) and low-power 3D systems-on-chip (SoC) is to guarantee system-wide power and signal integrity. This chapter provides an overview on interconnects with emphasis on through-silicon-via (TSV)-based 3D ICs. Different TSV types and their implications to power/signal integrity are first discussed. Power distribution methodology for a nine-plane processor-memory stack follows. Compact models are also proposed to achieve efficient noise coupling analysis in 3D ICs.

A 3D structure is an effective platform for integrating heterogeneous circuits within a single system. Each layer of a 3D IC is typically independently fabricated using different substrate materials for different applications [3]. The 3D structure is therefore dedicated to modern diverse applications as high-performance computing and mobile and wearable devices.

Although 3D ICs potentially offer increased system integration over 2D planar ICs, their integration densities vary for different technologies [2,9]. At the coarsest level, wire bonding may be used to connect multiple separately fabricated layers through connections at the periphery; for such technologies, the vertical interconnect pitch is of the order of 35 μm. A larger number of inter-layer connections are made possible by microbump technology, which uses bumps on die surface for vertical interconnect and can achieve smaller pitches of the order of 10 μm. The TSV approach can reduce the vertical pitch down to 1–5 μm by etching a hole right through the wafer to build a via that connects the layers. At the finest level, monolithic 3D integration can make 100 nm pitch vertical interconnections possible and enables the 3D stacking of transistors on the same wafer. It is generally accepted that true 3D integration is based on the use of TSVs or monolithic integration, and other technologies are sometimes lumped into the category of 2.5D integration.

Advancements in the field of VLSI have led to more compact ICs having higher clock frequencies and lower power consumptions. The paradigms of these integrated technologies are SoC and Systems-in-Package (SiP). Nowadays, these conventional 2D planar technologies begin to face several challenges from technological and financial points of view: physical limits, processing complexity, fabrication costs, etc. Consequently, technological approaches other than scaling are now investigated to continue following or get over Moore's Law. Novel 3D integration technology appears to be the most promising candidate. Currently, 3D integrated systems are obtained by stacking vertically 2D ICs made using mature, controlled micro, and nano technologies. The electrical connections are ensured by new pre- or post-processed metallic structures: redistribution layers (RDL), which distribute power and high-speed signals on die top or backside surfaces; copper pillars; and mostly TSV, which is a key enabling technology for 3D integration, propagating signals through the silicon layers.

Chips can also be stacked vertically using wire-bonding and flip-chip techniques. However, the flip-chip technique solely provides an interconnection between two chips whereas wire-bonding only enables the connecting of chip input/output pads located at their perimeter. Moreover, wire-bonding presents disadvantages in terms of surface and propagation delays depending on the application frequency clock and the wire-bond lengths. Finally, solder balls are put on the 3D systems' bottom layers' backsides to ensure connection with their environment. Compared with 2D classical schemes, 3D integration potential benefits are numerous, such as performance improvement and flexible heterogeneous system combinations (logic CMOS, RF analog function, memories, sensors). Stacking ICs vertically enables the significant shortening of the interconnect line networks, thereby decreasing interconnect line delays, which become a significant obstacle in 2D VLSI systems relative to delays in transistor switching, and the power dissipation while increasing the integration density, leading to smaller form factors and reduced fabrication costs [1,2].

Many challenges are encountered with the 3D integration due to its emerging technology status, notably in properly characterizing and electrically modeling the 3D interconnects. Additionally, few present-day CAD tools can design 3D architectures. Work has been mainly dedicated to advanced TSV electrical modeling as it plays a central role in realizing high-density integrated 3D systems and because it enables a large variation of shape, dimension (radius, length), dielectric thickness, and fill material. For example, reference [55–57] investigates the impacts of TSV cross-section and radius on frequency responses. References [58–60] gives some analytical expressions of the resistance, the self- and the mutual inductances of a copper-filled, tapered, high-density TSV; these are obtained from the cylindrical conductor formulas and by involving the current density. The case of two coupled cylindrical TSVs is studied in [61,62] and the closed-form expressions for the self and mutual parameters are reported; yet the two TSVs are isolated from their environment.

Indeed, the main difficulty in the TSV modeling is that the overall 3D interconnection electrical context must be considered, for example modeling the current paths if the environment comprises top/back redistribution metal lines. The parasitic coupling capacitance is analyzed in various configurations in [19,60] depending on whether the TSV is surrounded by top, bottom or side nearby interconnect wires. However, the TSV model in [19] was simplified by neglecting the TSV inductance, the oxide capacitances and by assuming a high-resistivity substrate. One other aspect of current interconnect modeling is that many authors describe their equivalent electrical models as lumped. When considering the global electrical context at low and medium frequencies, it is true that some interconnection elements of a network can still be described as lumped while their length is smaller than a tenth of the propagation signal wavelength; yet most of the interconnects electrical models must be distributed into several *RLCG* elementary cells to properly quantify the electromagnetic effects.

Finally, as with local metal layer interconnect line structures [20], the 3D interconnect global environment necessitates the modeling of the substrate coupling effects whose effects at high frequencies can no longer be neglected. Two different kinds of P-doped silicon substrate are typically used in CMOS/BiCMOS technology processes: uniform lightly doped substrates/high-resistive substrates (>1 Ω cm); and heavily doped substrates (<0, 1 Ω cm) with thin, lightly doped (>1 Ω cm) epitaxial layers [20]. Electrical modeling methods are then dependent on the application substrate type. Several substrate extraction techniques have been proposed in the literaturenb, but most of them cannot be employed for SoC with realistic dimensions or are only suitable for a particular type of structure or technology process [19,20].

In this chapter, the compact models of some 3D interconnects are presented, notably medium-density TSVs, reliable for low and medium frequencies (up to 10 GHz). They are derived from the Transmission Line Matrix method (TLM), 3D electromagnetic simulations, and parametrical extractions performed on various test structures. The efficient and simple modeling approach that we propose includes the consideration of the global electrical context and aims at providing rapidly information relative to signal integrity problems.

3D interconnects necessitate the consideration of the global electrical context to correctly evaluate system performances. The compact models must hence be derived from semiconductor and electromagnetic theory instead of being developed and extracted outside of a realistic 3D IC system environment; the electrical models could be restrictive in this latter case by being only useful for specific context. That is why the basic compact models proposed in the present work are, in first instance, based on parametrical extractions performed on test structures. The *RLCG* parametrical extractions are performed through S-parameters measurements with a Vector Network Analyzer (VNA) for a frequency sweep from 70 kHz to 40 GHz. The Thru-Reflect-Line (TRL) de-embedding technique is applied to remove from the S-parameters measurements the contact and access lines effects, and thus obtain the test structures' actual RF behaviors [61].

3.2 Presentation of Some Structures

Two kinds of test structure used for the modeling approach are presented in this study. The first one is the Back-End-Of-Line (BEOL) coplanar waveguide (CPW), which is used as a redistribution metal line to propagate electrical signals along the top or the backside of the stacked dies. The second one, which we will denote hereafter as TSV chain or nxTSV chain, consists in n medium-density TSVs connected at the backside of the chip with a Back RDL (BRDL) redistribution line. The TSVs are connected to the BEOL level, where the ground lines are located, by means of square contact pads. The physical views of a coplanar waveguide and a chain of 2 TSVs (2xTSV chain) are represented in Figure 3.1.

The test structures are realized using a highly conductive silicon substrate having a thickness of 120 µm, a resistivity ρ = 10 mΩ cm, and a relative permittivity ε_r = 11.7. On the substrate, a thin epitaxial layer is deposited (thickness: 4.5 µm, resistivity: 10 Ω cm, relative permittivity: 11.7). The lines at the BEOL level, for CPWs and TSV chains, are isolated from the substrate by an oxide layer (thickness: 0.8 µm, relative permittivity: 6.2). Regarding the TSV chains, an oxide layer (thickness: 1 µm, relative permittivity: 6.5) is added at the backside of the structures to isolate the BRDL line from the substrate. All the layers used for the CPW and the TSV chain structures are represented in Figure 3.2, with their respective characteristics. For the BEOL level lines have a thickness of 0.8 µm, whereas they are 7 µm for the RDL level.

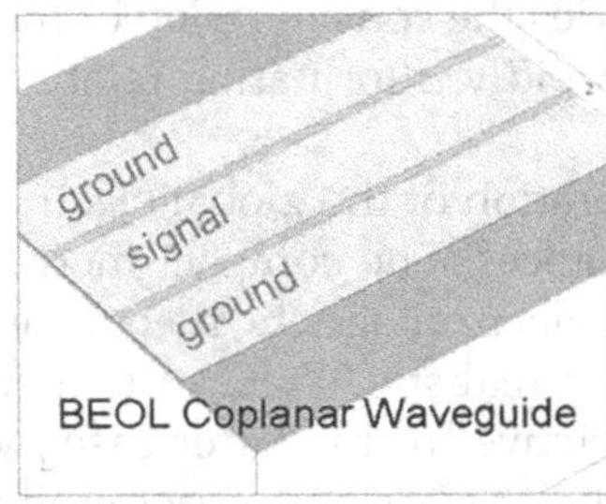

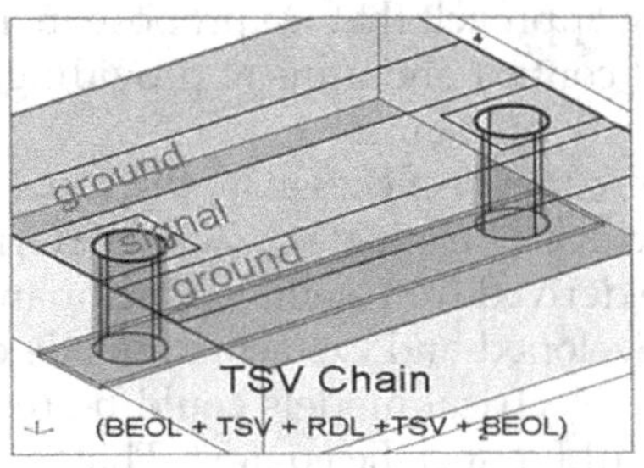

FIGURE 3.1
Physical views of some test structures. Left: a BEOL coplanar waveguide. Right: a chain of 2 TSVs (2xTSV chain).

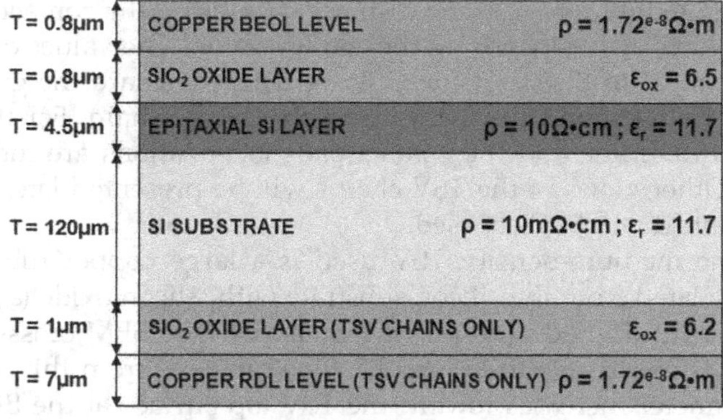

FIGURE 3.2
Geometrical and technological data of the layers comprised in the CPW and TSV chain structures.

The CPW structures are made with three copper lines. Two configurations, called W1 and W2, are considered for these structures. For the W1 configuration, the width of the signal line is equal to 90 µm (60 µm for W2) with a spacing between lines of 15 µm (30 µm for W2); the width of the ground lines equals 80 µm whatever the configuration. Three different coplanar line lengths have been investigated for W1 and W2 configurations; the values are: 60, 240 and 540 µm. A top view of the CPW is presented in Figure 3.3 with all the geometrical data indicated with the exception of the thickness.

For the TSV chains, the BRDL line length can have one of three values: 60, 240, and 540 µm. As with the CPWs, there are two configurations for the TSV chain. In the W1 configuration, the width of the BRDL line and the contact

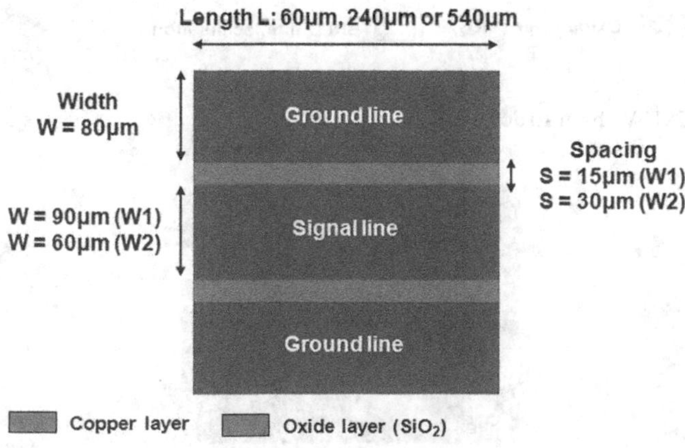

FIGURE 3.3
Top view of a CPW structure.

pads is 90 µm (60 µm for W2). The distance separating the contact pads and the ground lines also depends on the configuration. The values of the spacings are identical to those for the CPW structures; as are the ground line widths. Figure 3.4 shows the top view of the 2xTSV chain. For the sake of clarity, the BRDL line and the contact pads localizations are indicated on the figure. Other views of the TSV chains will be presented later when the electrical context will be discussed.

Finally, the medium-density TSV used is a large copper tubular conic structure, isolated from the silicon substrate with silicon oxide ($\varepsilon_{ox} = 6.5$) on the TSV walls and filled with polymer material. The SEM cross-section of such a TSV is observable in Figure 3.5. One can note from this figure that the TSV diameter increases toward the TSV top surface at the BEOL level, whereas the thicknesses of the copper and oxide layers decrease. For the TSV

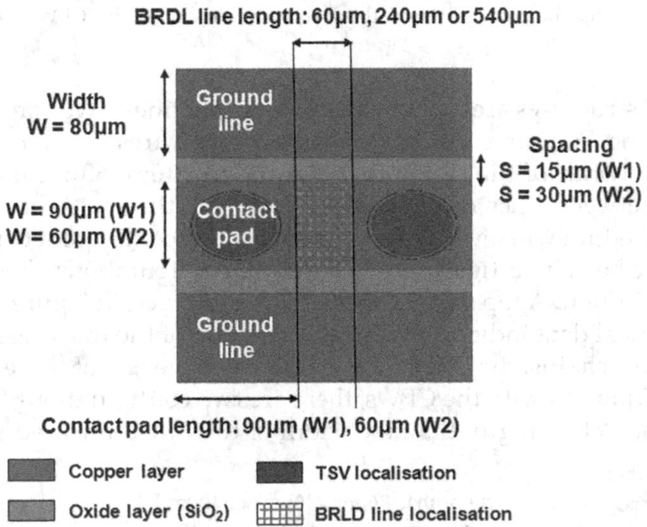

FIGURE 3.4
Top view of a 2xTSV chain structure

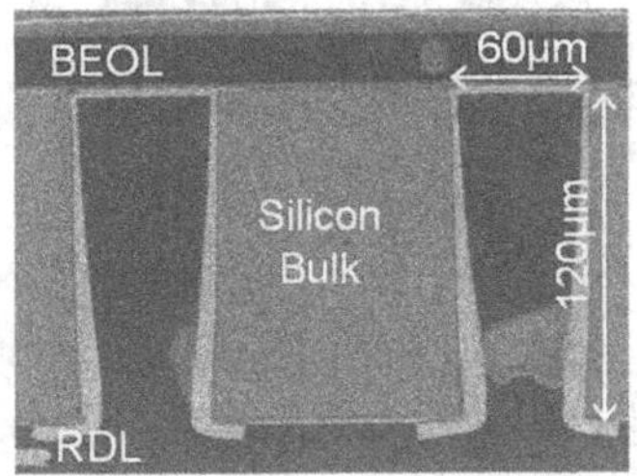

FIGURE 3.5
SEM cross section of the medium-density TSV used in the TSV chains (W1 configuration).

TABLE 3.1

Medium-density TSV Geometrical Characteristics (W1 and W2 Configurations)

TSV Geometrical Variation	BEOL Level	RDL Level
TSV diameter	60 μm (W1)	59 μm (W1)
(polymer material + copper layer)	40 μm (W2)	37 μm (W2)
TSV copper layer thickness	1 μm	3 μm
TSV oxide layer thickness	0.2 μm	0.5 μm

chain W1 configuration, the TSV diameter, including the copper layer and the polymer material, varies from 60 to 59 μm (from 40 to 37 μm for W2). For both configurations, the TSV copper and SiO_2 layers thicknesses vary at each extremity from 1 to 3 μm and from 0.2 to 0.5 μm, respectively (Table 3.1). All the TSV fabrication process details are reported in [19].

3.3 Compact Models of the Medium-TSV and the Coplanar Line

As in planar technologies, 3D interconnects can be built as *RLCG* equivalent electrical models described as Π or T networks. It does not matter which type of network as long as the interconnect length is smaller than the tenth of the propagated signal wavelength. Otherwise, the two networks do not have the same RF behavior and they must be distributed in a certain number of elementary cells to give equivalent responses. The T network will be considered later on in this work. Each of the 3D interconnect *RLCG* networks is modeled with serial elements (resistances or partial inductances) to model the signal propagation and parallel elements (capacitances or conductances) to model the interconnect environment. The proximity (or coupling) effects are also included in the compact model description. The geometrical parameters describing the analytical expressions are all expressed in meters. The coplanar line resistances and the self and mutual inductances are calculated from well-known formulas (3.1–3.3). The resistances are calculated for a DC value (the skin effect could be considered later) and the inductances are calculated depending on the partial inductance analytical expressions [59].

$$R = \rho \frac{L}{W \cdot T} \tag{3.1}$$

$$L_{self} = \mu_0 \mu_r \frac{L}{2\pi}\left[\ln\left(\frac{2L}{W+T}\right) + 0.5 + \frac{0.447(W+T)}{2L}\right] \tag{3.2}$$

$$M = \mu_0 \mu_r \frac{L}{2\pi}\left[\ln\left(\frac{2L}{p}\right) - 1 + \frac{p}{L}\right] \tag{3.3}$$

where ρ is the metal resistivity (copper: $1.72 \cdot 10^{-8}$ Ω m); μ_0 and μ_r are the vacuum and the copper relative permeabilities; and *W*, *L*, *T*, and *P* are, respectively, the line width, length, thickness and the pitch between two interconnect lines (Figure 3.6).

The interline capacitance C_{inter} (or coupling capacitance) is calculated according to the set of equations defined in [59] from the concerned line surfaces, and the fringe capacitances taken at the extremities (*Cf*) and the middle (*Cf'*) of the coupled-line system. The electrical parameter *Cp* corresponds to the coupling capacitance between the ground plane and the line surface concerned (Figure 3.7); it is also referred to the literature as the line's self-capacitance.

$$Cf = \varepsilon_0 \varepsilon_{ox}\left(0.075\frac{W}{H} + 1.4\left(\frac{T}{H}\right)^{0.222}\right) \tag{3.4}$$

$$Cf' = \frac{Cf}{1 + (H/S)} \tag{3.5}$$

$$Cp = \varepsilon_0 \varepsilon_{ox} \frac{W \cdot L}{H} \tag{3.6}$$

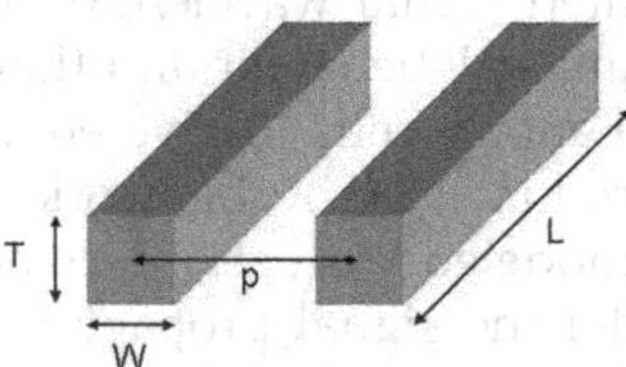

FIGURE 3.6
Schematic of the coplanar lines.

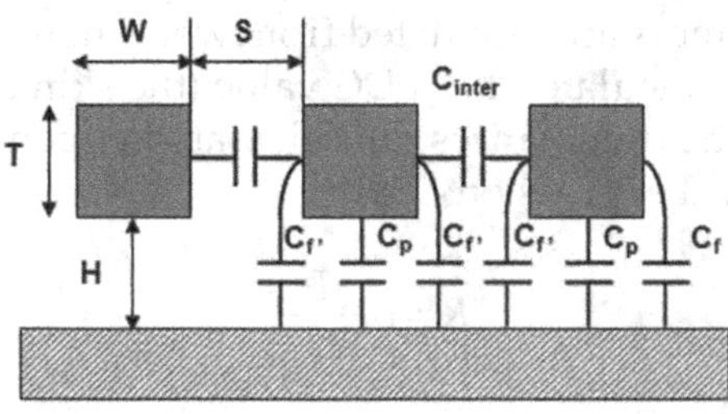

FIGURE 3.7
Illustration of the capacitive couplings between parallel lines located over a ground plane.

$$C_{\text{inter}} = Cf - Cf' + \frac{\varepsilon_0 \varepsilon_{ox} L}{100}\left(3\frac{W}{H} + 83\frac{T}{H} - 7\left(\frac{T}{H}\right)^{0.222}\right)\cdot\left(\frac{H}{S}\right)^{1.24} \tag{3.7}$$

where ε_{ox} and ε_0 are the silicon oxide and the vacuum relative permittivities, and S is the interline gap. An average value of the width W in formula (3.7) is used as signal and ground lines are geometrically different in width.

The electrical parameter values must be adapted if the compact model is to depict a lumped model or an elementary cell of a distributed model. The coplanar waveguide model is illustrated in Figure 3.1 with all the lines connected to the substrate through a node SUB. Depending on the system geometrical and technological data, the inductive coupling between the two ground lines can be neglected or not.

In the TSV chain, the currents are flowing vertically through the epitaxial and the oxide layers to reach the ground planes corresponding to the BEOL level ground lines. The electrical parameters of the epitaxial and oxide layers ($R_{epi(TSV)}$, $C_{epi(TSV)}$ and $C_{ox(GND)}$) are calculated in the same way as the coplanar waveguide using expressions (3.6). The coupling between the substrate and the TSVs is taken into account with the geometrical capacitance $C_{ox(TSV)}$ (here the silicon depletion region capacitance is neglected as it introduces minor effect due to the highly doped substrate. Only the proximity effects between TSVs (TSV/TSV couplings) have been taken into account. The TSV chain electrical context modeling involves dividing the BEOL ground lines into serial blocks, by dissociating the parts reached on the overall surface by the current coming from the TSVs. These surface parts have a length corresponding to that of the TSVs contact pads L_{pad}. The width and the thickness remain unchanged for the entire line surface. Hence the epitaxial layer parameters ($R_{epi(TSV)}$, $C_{epi(TSV)}$), as well as the oxide capacitance ($C_{ox(GND)}$), modeling the current path going through the TSV to the ground line, must be calculated for $L = L_{pad}$.

The parts of the surfaces of the ground lines involved in the current paths are simply modeled by RL networks; otherwise they are modeled, for instance, as two ground lines decomposed into serial blocks in the case of a chain of two TSVs (2xTSV chain—see below). From the way the TSVs are connected, the 2xTSV chain can be seen as a "U-shaped" structure. For nxTSV chains, $n/2$ "U-shaped" structures are connected at the BEOL level by contact pads and a 60 μm length access line, whatever the configuration. The BEOL ground line total length thus equals to:

$$L_{\text{BEOL}} = nL_{\text{pads}} + \frac{n}{2}L_{\text{space}} + \left(\frac{n}{2} - 1\right)L_{\text{access}} \tag{3.8}$$

The efficient and simple modeling approach is based on subdividing the whole 3D interconnect path in different blocks; simple analytical formulas are used to build the global model. Figure 3.7 illustrates the ground lines decomposition into serials blocks in the case of the 4xTSV chain. The decomposition can be extended for chains including a higher number of TSVs.

3.3.1 3D Transmission Line Extractor 3D-TLE

Because system descriptions can become complex when considering the global electrical context and the compact models' distribution, we have developed using MATLAB® [63] a 3D extraction tool, 3D-TLE (3D Transmission Line Extractor), based on our modeling approach and which integrates the substrate extraction method algorithm (Figure 3.8). Through a hierarchical syntax and specific statements, the user depicts in a text file the geometrical and technological data of his/her system by defining: its layers; the type of components comprised in the system; the connectivity between chip elements, which are instantiated components and the element interactions (couplings, interactions with layers for current path modeling).

The extraction tool generates from this text file a SPICE netlist, containing the system *RLCG* electrical description, exportable toward CAD tools such as Agilent Technologies ADS® [64]. The viability of 3D-TLE has been checked through S-parameter comparisons for given systems between their 3D-TLE SPICE netlists and their schematics designed under ADS®. Obviously, the expected results show that the responses are very closed; however, describing the structure using 3D-TLE is easier and faster than designing it using the CAD tool. 3D-TLE provides also the system 3D picture.

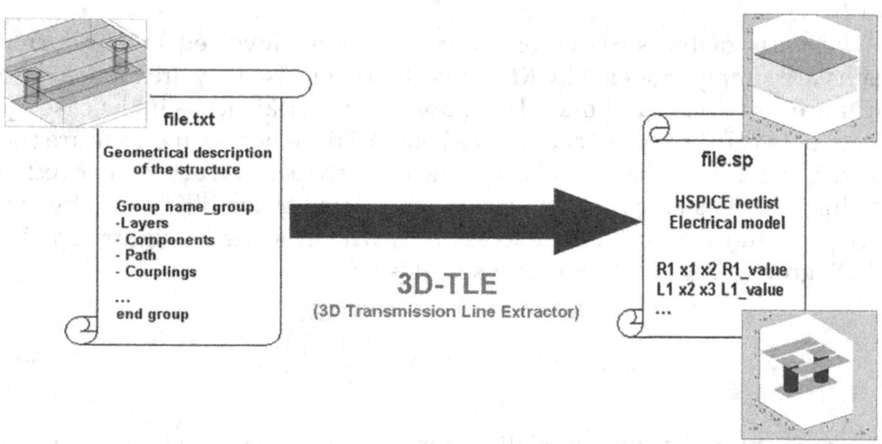

FIGURE 3.8
In-house 3D extraction tool 3D-TLE (3D Transmission Line Extractor) environment.

3.3.2 Modeling Approach Validation and Test Structures' RF Behaviors

The modeling approach is validated in the frequency domain via S-parameter comparisons between the measurements performed on the test structures, for both W1 and W2 configurations, and the simulation results from their equivalent electrical models. Some of these comparison results are presented in Figures 3.9–3.12. The test structures' RF responses are plotted with dotted blue curves; those of the electrical models with plain red curves. The equivalent *RLCG* compact models proposed for the coplanar waveguides and the TSV chains demonstrate very good accuracy for frequencies up to 20 and 10 GHz, respectively. Some spikes can be observed in measurements but not in simulations at low frequency (below 1 GHz). These resonances are due to the coupling between the test structure substrate and its measurement environment. The test sample chuck is a piece of metal which acts as an excellent antenna with uncontrolled impedance to the measurement ground. This unknown impedance explains the unexpected poles of resonances. The spikes

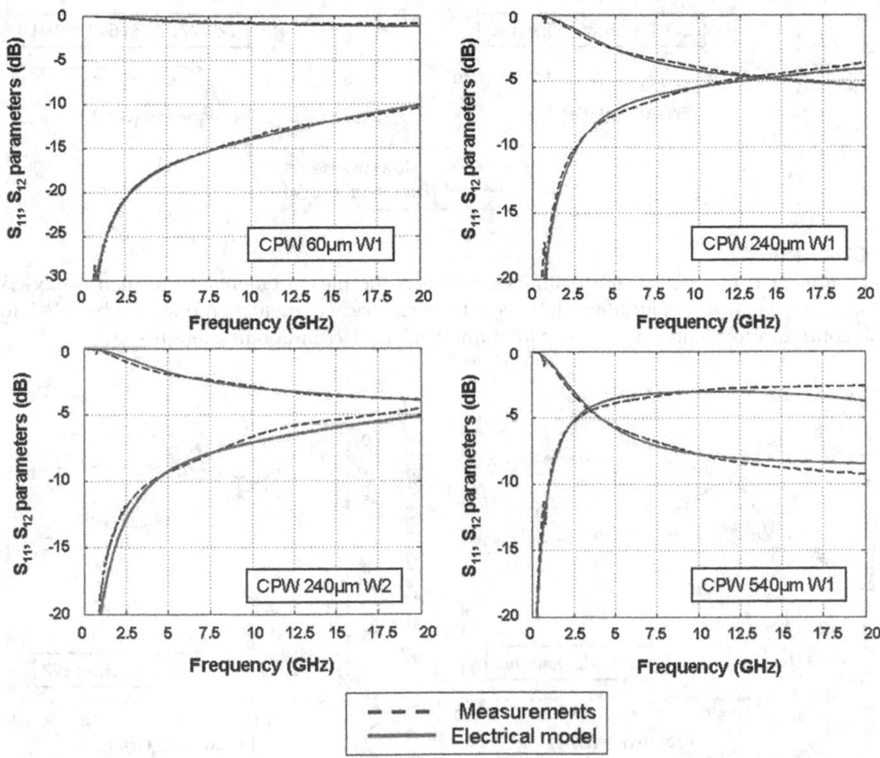

FIGURE 3.9
S-parameter comparisons, up to 20 GHz, between the measurements performed on coplanar waveguides test structures and the simulations of their electrical equivalent models for both W1 and W2 configurations and different coplanar lines lengths.

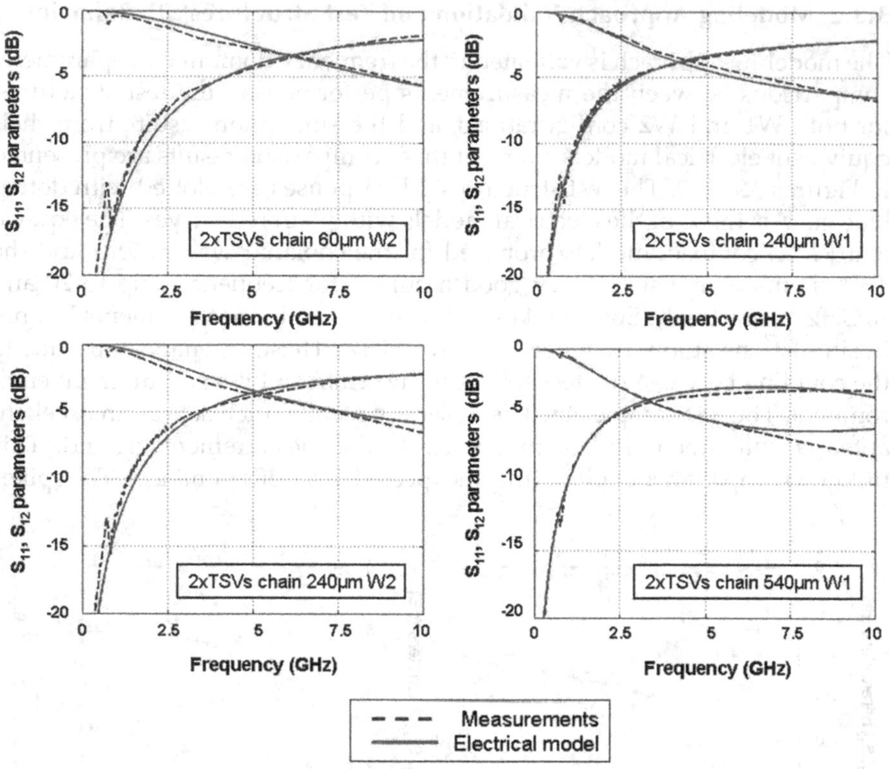

FIGURE 3.10
S-parameter comparisons, up to 10 GHz, between the measurements performed on 2xTSVs chain test structures and the simulations of their electrical equivalent models for both W1 and W2 configurations and different coplanar and back redistribution line lengths.

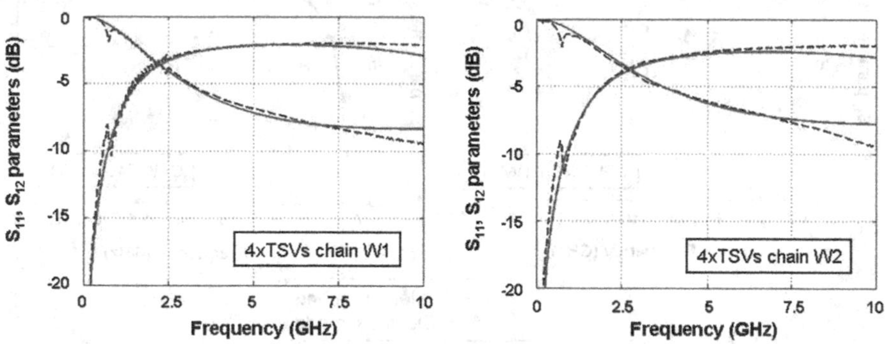

FIGURE 3.11
S-parameter comparisons, up to 10 GHz, between the measurements performed on 4xTSVs chain test structures and the simulations of their electrical equivalent models for both W1 and W2 configurations.

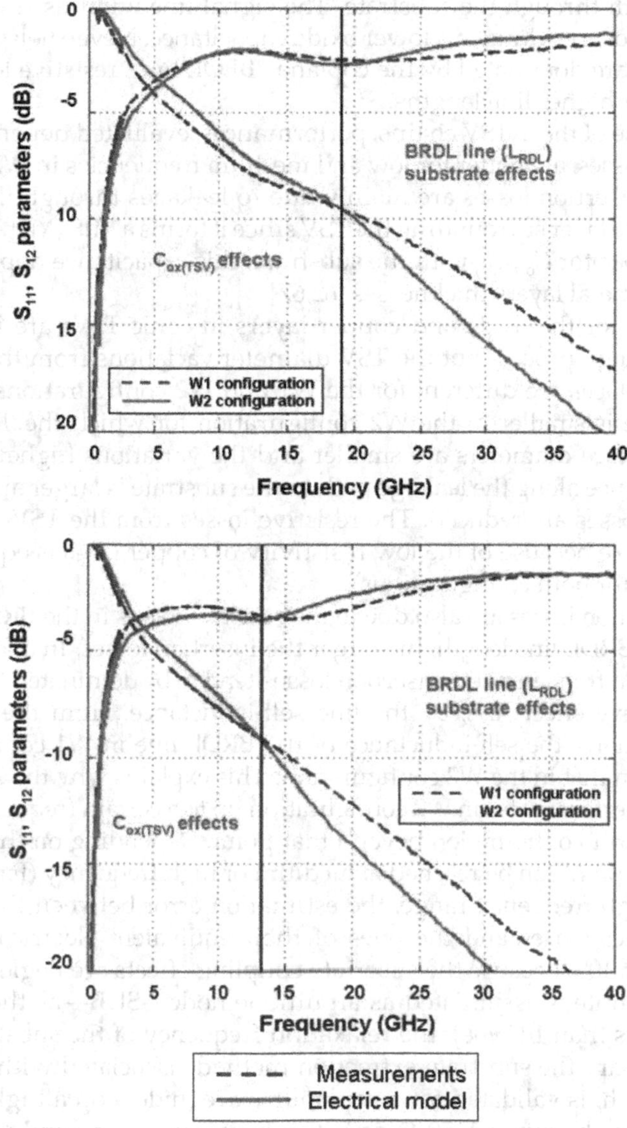

FIGURE 3.12
S-parameter comparisons for 2xTSVs chains between W1 and W2 configurations. Top: the BRDL line length is 420 µm; the spacing between contact pads at the BEOL level is 240 µm. Bottom: the BRDL line length is 720 µm; the spacing between contact pads at the BEOL level is 540µm.

can be suppressed by isolating the test structure from its environment, implying the substrate has to be strongly grounded or biased.

The coplanar waveguides in W2 configuration give rise to better transmission resulting from less insertion losses due to higher impedance along the leakage path through the substrate. The signal line width is smaller in this configuration, resulting in a lower oxide capacitance. Nevertheless, the insertion losses are dominated by the coplanar BEOL lines resistive losses which increase for higher line lengths.

In the case of the 2xTSV chains, performances, evaluated depending on the insertion losses, are better for low and medium frequencies in W2 configuration. The insertion losses are mainly due to leakages through the substrate via the oxide layer surrounding the TSV since it forms a MIS (Metal-Insulator-Silicon) capacitor $C_{ox(TSV)}$ with the substrate; this capacitance depends on the oxide and metal layers thicknesses [62,67].

The data for the oxide and copper layers in conic TSVs are imposed by the technology process but the TSV diameter variations from the top to the bottom surfaces are different for the W1 and W2 configurations. The oxide capacitance is smaller in the W2 configuration for which the TSV top and bottom surface diameters are smaller and the variations higher. Therefore, the impedance along the leakage path to the substrate is larger and hence the insertion losses are reduced. The resistive losses from the TSVs themselves are negligible because of the low resistivity of copper (R_{TSV} is equal to a few milliohms for both configurations).

The insertion losses are also due to the resistive losses in the BRDL line. The higher the BRDL line length, the larger the insertion losses. In addition, when increasing in frequency, the insertion losses tend to be dominated by the BRDL line inductive effects, i.e., by the line self-inductance. From their geometrical descriptions, the self-inductance of the BRDL line in W1 configuration is smaller than that in the W2 configuration. This explains why the 2xTSV chain has better performances in W2 configuration up to a certain frequency and better ones in W1 configuration beyond that point. Depending on the BRDL line length, this point can be reached at medium or high frequency (Figure 3.12).

In the high-frequency range, the estimation error between the test structures' RF responses and the ones of their equivalent electrical models is higher than 10% because the substrate coupling effects are neglected.

The substrate is assimilated as an unique node—SUB—as the frequency is much less than $1/(2\pi\rho\varepsilon)$, the relaxation frequency of the substrate). In the following part, the substrate extraction method, associated with the modeling approach, is validated for a coplanar waveguide atop a highly resistive substrate (in the range 1–50 Ω cm). The S-parameters given by the electrical model (with or without substrate model) have been compared, for a frequency range up to 10 GHz, with the experimental data. The results of this comparison are shown in Figure 3.13.

The results presented in this section illustrate our efficient and simple modeling approach applied to quite complex 3D interconnects (e.g., TSV chains). Such pragmatic approach presents the advantage to be automatized for large circuit extraction (an example of TSV matrix is described in below). However, the method can be enriched by including skin effect and eddy current, leading to huge complexity for the automatic extraction.

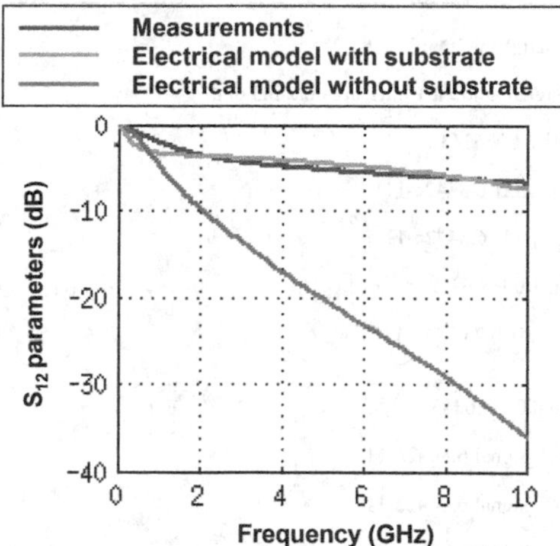

FIGURE 3.13
S-parameter comparisons for a 3800 μm coplanar waveguide system atop a highly resistive substrate (1–50 Ω cm). Left: S_{11} parameters. Right: S_{12} parameters.

When getting the SPICE-compatible file from 3D-TLE, one can import it to ADS easily.

An example of the generated SPICE txt file is shown in Figure 3.14. Table 3.2 shows the different physical parameters.

If we change only the substrate, in ADS, we can establish a changeable substrate circuit (Figure 3.15).

The result can be seen in Figure 3.16.

As we can see, if we change only the substrate conductivity, it will not match with the measurement. So we can see that the substrate conductivity has a major effect on the high-frequency transmission. The higher the substrate conductivity, the more obvious gets the signal attenuation (Figure 3.17).

3.3.2.1 High-Resistive Substrate Study

S-parameters for CPW model of high-resistive substrate, in dB, are shown in Figure 3.17.

Table 3.3 shows comparison between original parameters, optimized parameters, and 3D-TLE for high-resistive substrate.

3.3.2.2 Transient Study—Eye Diagram

The eye diagram can be used to evaluate the extracted interconnection transmission's quality (3D-TLM). In order to simplify the simulation, we input a numeral signal and measure the output signal. The output signal is more or

```
*Netlist generated from 3D_IDEAS
.subckt gnd1_signal_gnd2 in1 out1 in2 out2 in3 out3
R1 in1 x1_gnd1 0.0066343
L1 x1_gnd1 x2_gnd1 6.8442e-11
L2 x2_gnd1 x3_gnd1 6.8442e-11
R2 x3_gnd1 out1 0.0066343
Cox_gnd1 x2_gnd1 0 2.0741e-13

R3 in2 x1_signal 0.0066343
L3 x1_signal x2_signal 6.8442e-11
L4 x2_signal x3_signal 6.8442e-11
R4 x3_signal out2 0.0066343
Cox_signal x2_signal 0 2.0741e-13

R5 in3 x1_gnd2 0.0066343
L5 x1_gnd2 x2_gnd2 6.8442e-11
L6 x2_gnd2 x3_gnd2 6.8442e-11
R6 x3_gnd2 out3 0.0066343
Cox_gnd2 x2_gnd2 0 2.0741e-13

K1 L1 L3 0.41668
K2 L2 L4 0.41668
Cinter1 x2_signal x2_gnd1 2.4691e-15

K3 L3 L5 0.41668
K4 L4 L6 0.41668
Cinter2 x2_gnd2 x2_signal 2.4692e-15

K5 L1 L5 0.28941
K6 L2 L6 0.28941

.ends gnd1_signal_gnd2

X1_gnd1_signal_gnd2 0 1 2 3 0 4 gnd1_signal_gnd2

.subckt pillier1_pillier2_pillier3 in1 out1 in2 out2 in3 out3
```

FIGURE 3.14
Exportable SPICE-compatible TXT file.

(*Continued*)

```
R1 in1 x1_pillier1 0.000219
L1 x1_pillier1 x2_pillier1 5.3753e-12
L2 x2_pillier1 x3_pillier1 5.3753e-12
R2 x3_pillier1 out1 0.000219

R3 in2 x1_pillier2 0.000219
L3 x1_pillier2 x2_pillier2 5.3753e-12
L4 x2_pillier2 x3_pillier2 5.3753e-12
R4 x3_pillier2 out2 0.000219

R5 in3 x1_pillier3 0.000219
L5 x1_pillier3 x2_pillier3 5.3753e-12
L6 x2_pillier3 x3_pillier3 5.3753e-12
R6 x3_pillier3 out3 0.000219

K1 L1 L3 0.22791
K2 L2 L4 0.22791
Cinter1 x2_pillier2 x2_pillier1 1.0556e-15

K3 L3 L5 0.22791
K4 L4 L6 0.22791
Cinter2 x2_pillier3 x2_pillier2 1.0556e-15

K5 L1 L5 0.11562
K6 L2 L6 0.11562

.ends pillier1_pillier2_pillier3

X1_pillier1_pillier2_pillier3 1 10 3 30 4 40 pillier1_pillier2_pillier3

.subckt gauche_centre_droite in1 out1 in2 out2 in3 out3

R1 in1 x1_gauche 0.013269
L1 x1_gauche x2_gauche 1.7262e-10
L2 x2_gauche x3_gauche 1.7262e-10
R2 x3_gauche out1 0.013269
Cox_gauche x2_gauche 0 2.7599e-13
```

FIGURE 3.14 (CONTINUED)
Exportable SPICE-compatible TXT file.

(Continued)

```
R3 in2 x1_centre 0.013269
L3 x1_centre x2_centre 1.7262e-10
L4 x2_centre x3_centre 1.7262e-10
R4 x3_centre out2 0.013269
Cox_centre x2_centre 0 2.7599e-13
R5 in3 x1_droite 0.013269
L5 x1_droite x2_droite 1.7262e-10
L6 x2_droite x3_droite 1.7262e-10
R6 x3_droite out3 0.013269
Cox_droite x2_droite 0 2.7599e-13
K1 L1 L3 0.48924
K2 L2 L4 0.48924
Cinter1 x2_centre x2_gauche 5.2569e-15
K3 L3 L5 0.48924
K4 L4 L6 0.48924
Cinter2 x2_droite x2_centre 5.2569e-15
K5 L1 L5 0.33042
K6 L2 L6 0.33042
.ends gauche_centre_droite
X1_gauche_centre_droite 10 11 30 31 40 41 gauche_centre_droite
.subckt pillier4_pillier5_pillier6 in1 out1 in2 out2 in3 out3
R1 in1 x1_pillier4 0.000219
L1 x1_pillier4 x2_pillier4 5.3753e-12
L2 x2_pillier4 x3_pillier4 5.3753e-12
R2 x3_pillier4 out1 0.000219
```

FIGURE 3.14 (CONTINUED)
Exportable SPICE-compatible TXT file.

(*Continued*)

```
R3 in2 x1_pillier5 0.000219
L3 x1_pillier5 x2_pillier5 5.3753e-12
L4 x2_pillier5 x3_pillier5 5.3753e-12
R4 x3_pillier5 out2 0.000219

R5 in3 x1_pillier6 0.000219
L5 x1_pillier6 x2_pillier6 5.3753e-12
L6 x2_pillier6 x3_pillier6 5.3753e-12
R6 x3_pillier6 out3 0.000219

K1 L1 L3 0.22791
K2 L2 L4 0.22791
Cinter1 x2_pillier5 x2_pillier4 1.0556e-15

K3 L3 L5 0.22791
K4 L4 L6 0.22791
Cinter2 x2_pillier6 x2_pillier5 1.0556e-15

K5 L1 L5 0.11562
K6 L2 L6 0.11562

.ends pillier4_pillier5_pillier6

X1_pillier4_pillier5_pillier6 11 12 31 32 41 42 pillier4_pillier5_pillier6

.subckt gnd1_end_signal_end_gnd2_end in1 out1 in2 out2 in3 out3

R1 in1 x1_gnd1_end 0.0066343
L1 x1_gnd1_end x2_gnd1_end 6.8442e-11
L2 x2_gnd1_end x3_gnd1_end 6.8442e-11
R2 x3_gnd1_end out1 0.0066343
Cox_gnd1_end x2_gnd1_end 0 2.0741e-13

R3 in2 x1_signal_end 0.0066343
L3 x1_signal_end x2_signal_end 6.8442e-11
L4 x2_signal_end x3_signal_end 6.8442e-11
R4 x3_signal_end out2 0.0066343
Cox_signal_end x2_signal_end 0 2.0741e-13
```

FIGURE 3.14 (CONTINUED)
Exportable SPICE-compatible TXT file.

(Continued)

```
R5 in3 x1_gnd2_end 0.0066343
L5 x1_gnd2_end x2_gnd2_end 6.8442e-11
L6 x2_gnd2_end x3_gnd2_end 6.8442e-11
R6 x3_gnd2_end out3 0.0066343
Cox_gnd2_end x2_gnd2_end 0 2.0741e-13

K1 L1 L3 0.41668
K2 L2 L4 0.41668
Cinter1 x2_signal_end x2_gnd1_end 2.4691e-15

K3 L3 L5 0.41668
K4 L4 L6 0.41668
Cinter2 x2_gnd2_end x2_signal_end 2.4692e-15

K5 L1 L5 0.28941
K6 L2 L6 0.28941
.ends gnd1_end_signal_end_gnd2_end
X1_gnd1_end_signal_end_gnd2_end 12 13 32 33 43 44 gnd1_end_signal_end_gnd2_end
.end
```

FIGURE 3.14 (CONTINUED)
Exportable SPICE-compatible TXT file.

TABLE 3.2
Restored Layers Physical Parameters

	Conductivity (Ω*m)	Relative Permittivity
SUBSTRATE	$1e^{-4}$	11.7
EPITAXIAL	0.1	13.5
OXIDE		11
BEOL	$2e^{-8}$	

less deformed according to the interconnection, which can be seen through a bigger or smaller eye diagram [68–71].

We can get the eye diagram by using ADS's <Eye Probe> and <Simulation-Transient> study. We put a bit sequential voltage source at the input by accompanying a 50 Ω resistance at its output. We attach the under test numeral component in the circuit and add a 100 fF capacitance at the output. We can get three different settings (Figures 3.18–3.20).

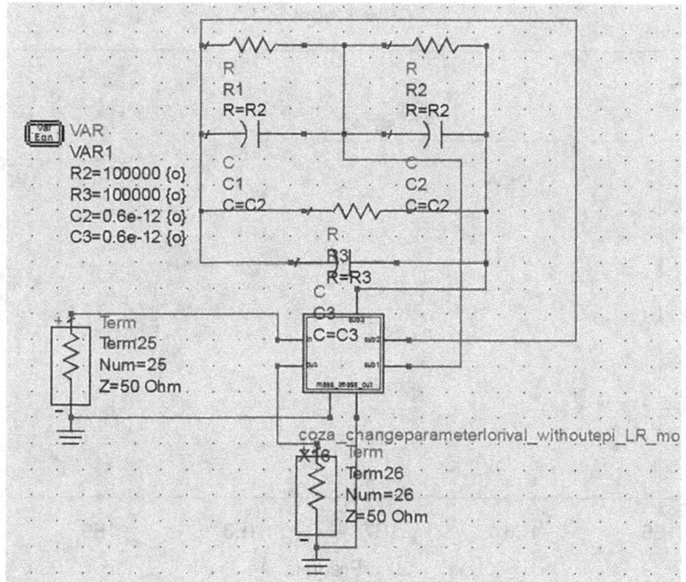

FIGURE 3.15
Importation in ADS of the extracted 3D-TLE CPW model.

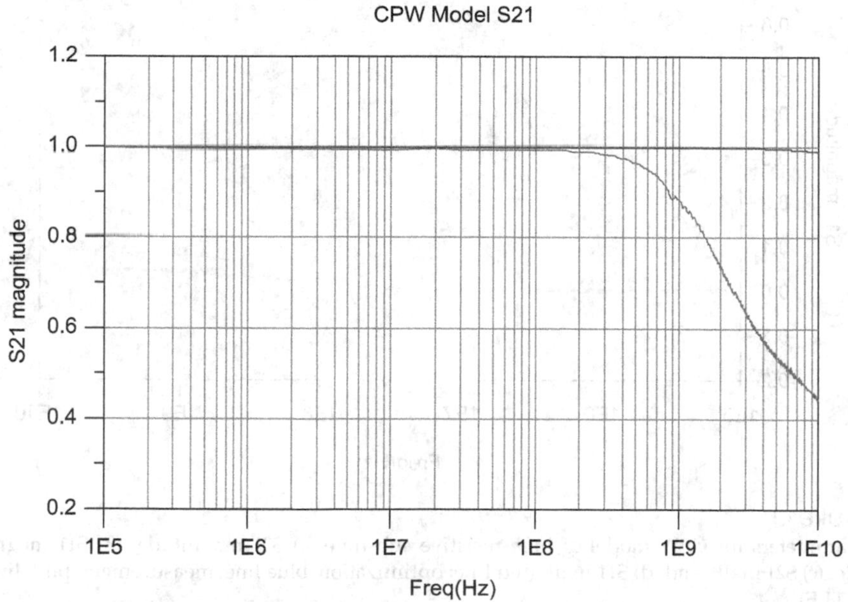

FIGURE 3.16
S21 magnitude by changing substrate conductivity (red line: optimization; blue line: measurement; pink line: 3D-TLE).

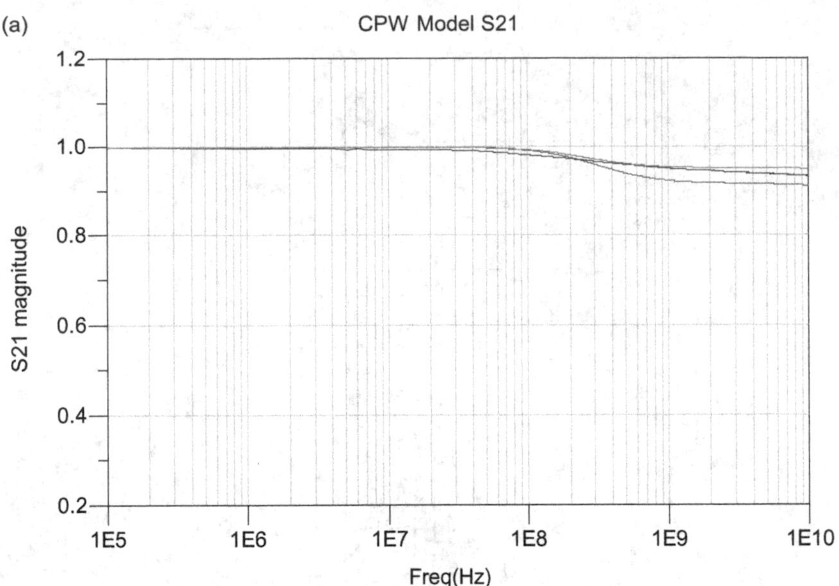

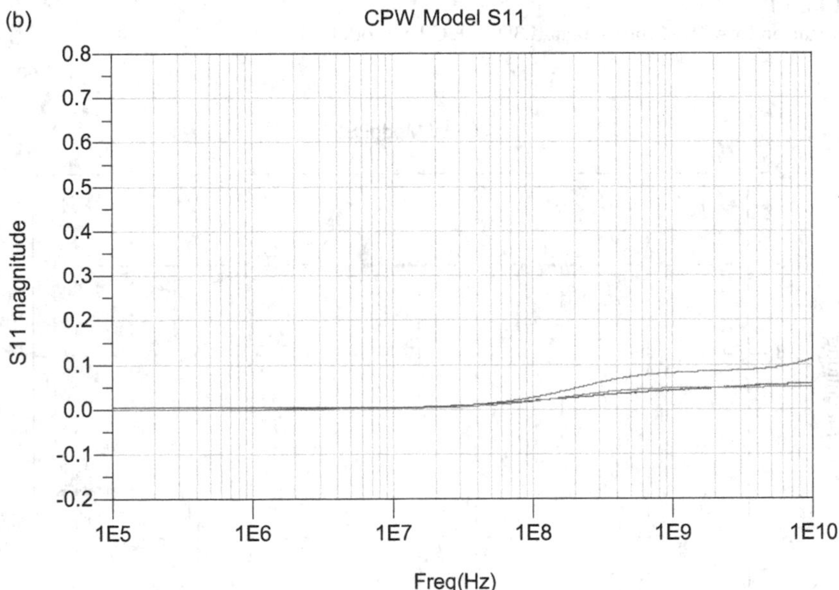

FIGURE 3.17
S-parameters for CPW model of high-resistive substrate (a) S21 magnitude, (b) S11 magnitude, (c) S21 in dB, and (d) S11 in dB (red line: optimization; blue line: measurement; pink line: 3D-TLE).

(Continued)

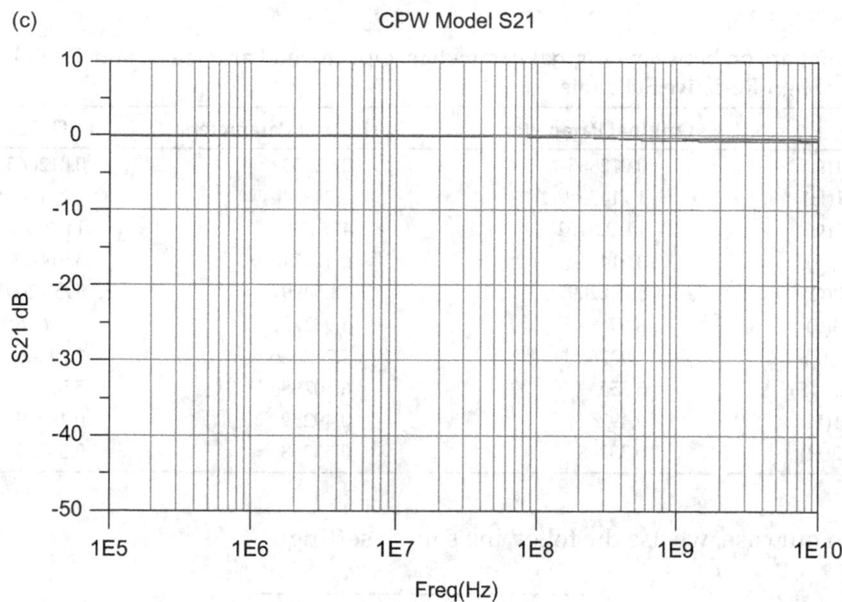

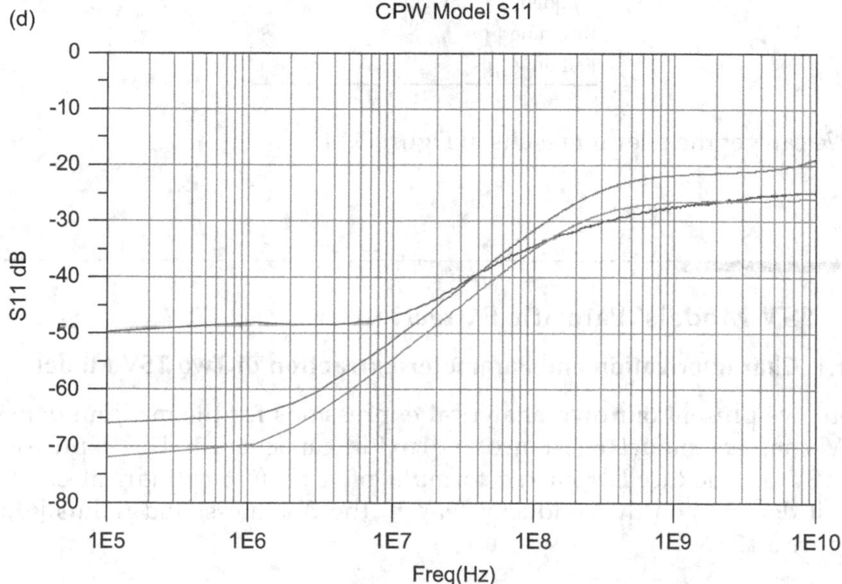

FIGURE 3.17 (CONTINUED)
S-parameters for CPW model of high-resistive substrate (a) S21 magnitude, (b) S11 magnitude, (c) S21 in dB, and (d) S11 in dB (red line: optimization; blue line: measurement; pink line: 3D-TLE).

TABLE 3.3
Comparison between Original Parameters, Optimized Parameters, and 3D-TLE for High-Resistive Substrate

	Original Parameter	Optimized Parameters	3D-TLE
R1(Ω)	0.012665	0.02533	0.012665
L1(H)	$1.7145e^{-10}$	$1.78984e^{-10}$	$1.7145e^{-10}$
R2(Ω)	1473.3402	478.838	1473.3402
C2(F)	$3.514e^{-14}$	$6.91474e^{-14}$	$3.514e^{-14}$
R3(Ω)	5953.2097	18.5399	5953.2097
C3(F)	$8.6966e^{-15}$	$4.95273e^{-17}$	$8.6966e^{-15}$
C_{ox}(F)	$1.9328e^{-12}$	$2.77256e^{-12}$	$1.9328e^{-12}$
C_{inter}(F)	$5.55e^{-15}$	$6.10788e^{-14}$	$5.55e^{-15}$
K1(H)	0.49259	0.49259	0.49259
K2(H)	0.33268	0.33268	0.33268

In our case, we use the following source setting:

$V_{low} = 0\,\text{V}$ and $V_{high} = 3.3\,\text{V}$
Frequency = 2.5 GHz
Rise time 1 ps
Fall time 1 ps

We can get the relevant results in Figure 3.21.

3.4 TSV Models' Parasitic Extraction

3.4.1 Characterization and Parameter Extraction of Two TSV Model

Here, we present our own analytical expressions for the medium-density TSV compact model resistance and oxide capacitance. The expressions of TSV are deduced from the formula of a perfect cylindrical conductor; it depends on the conductor length, the thickness, and radius [61,62] (Figure 3.22):

$$R_{ot} = R_{mt} + T_{ot} \tag{3.9}$$

$$R_{od} = R_{md} + T_{od} \tag{3.10}$$

$$R_{TSV} = \frac{H}{\pi\sigma} \frac{1}{2(R_{mt}T_{md} - R_{md}T_{mt})} \ln\left(\frac{T_{md}}{2(R_{md} - T_{md})} \cdot \frac{2R_{mt} - T_{mt}}{T_{mt}}\right) \tag{3.11}$$

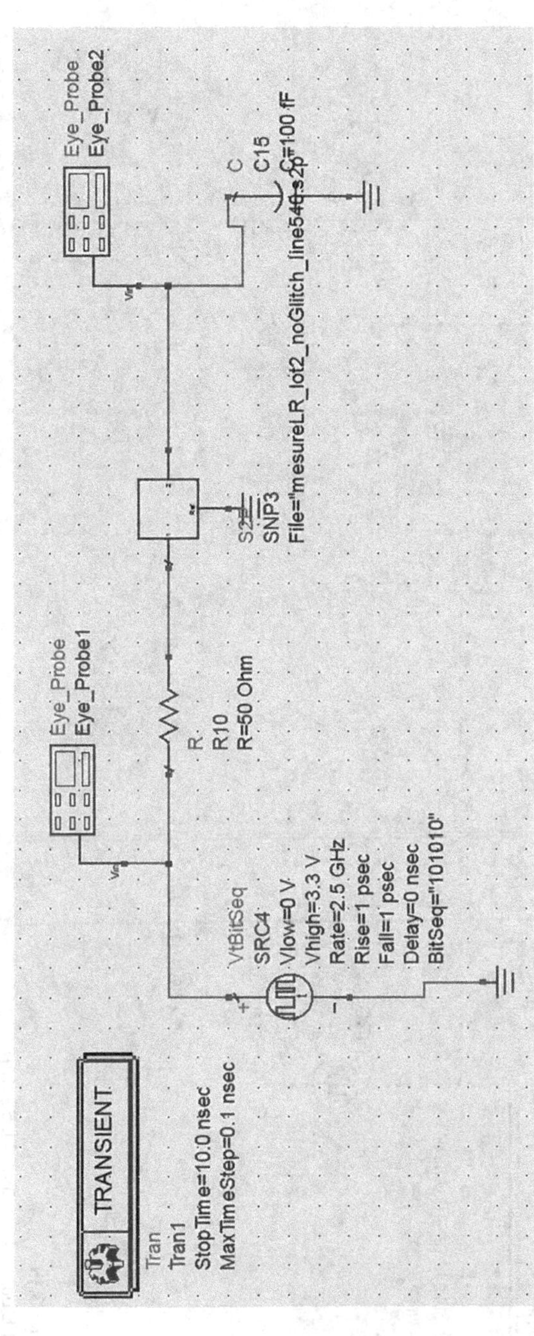

FIGURE 3.18
Measurements based on transient ADS setup.

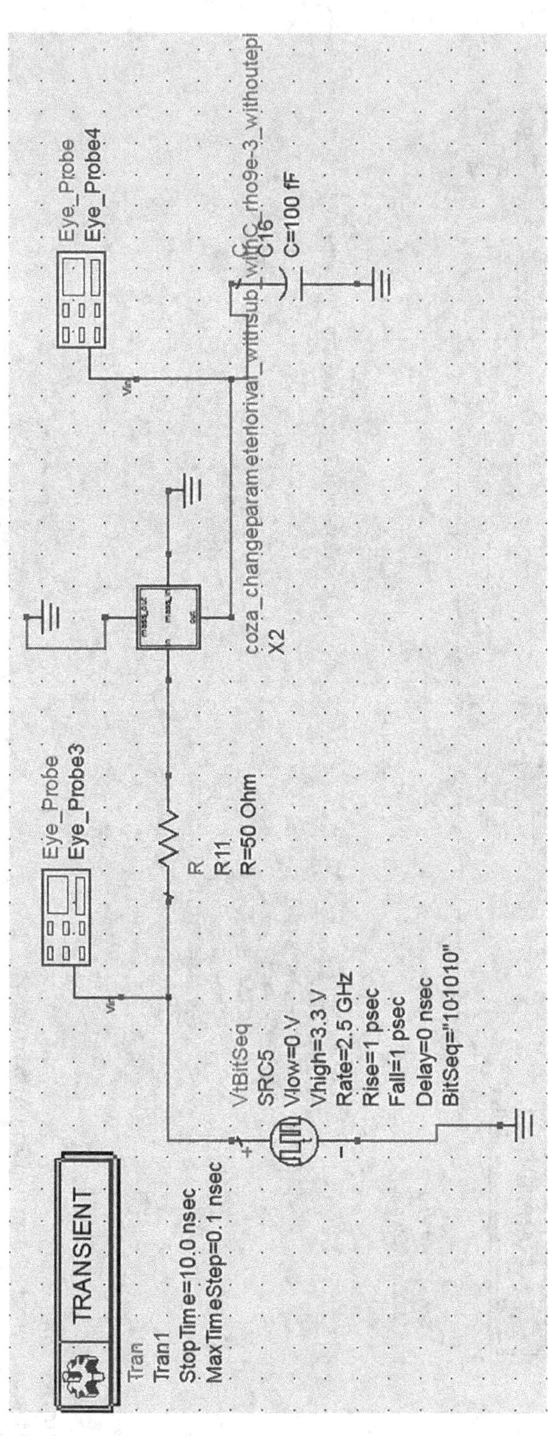

FIGURE 3.19
3D-TLE based transient ADS setup.

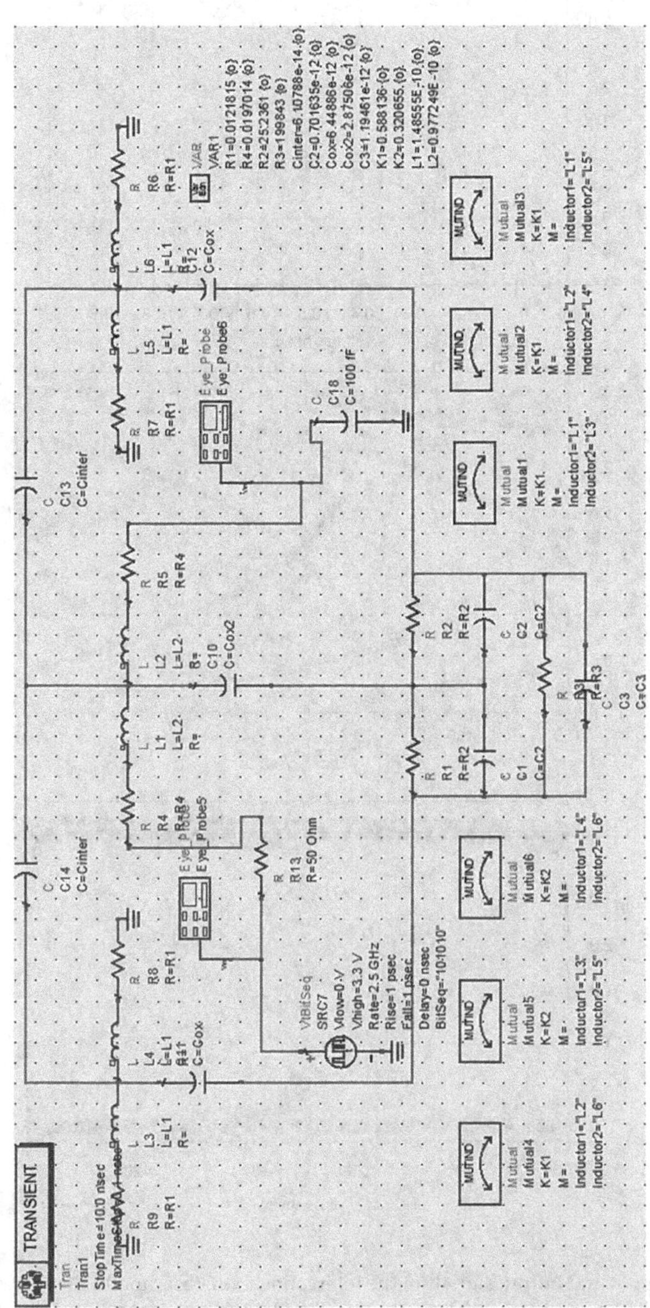

FIGURE 3.20
Transient Optimization Circuit.

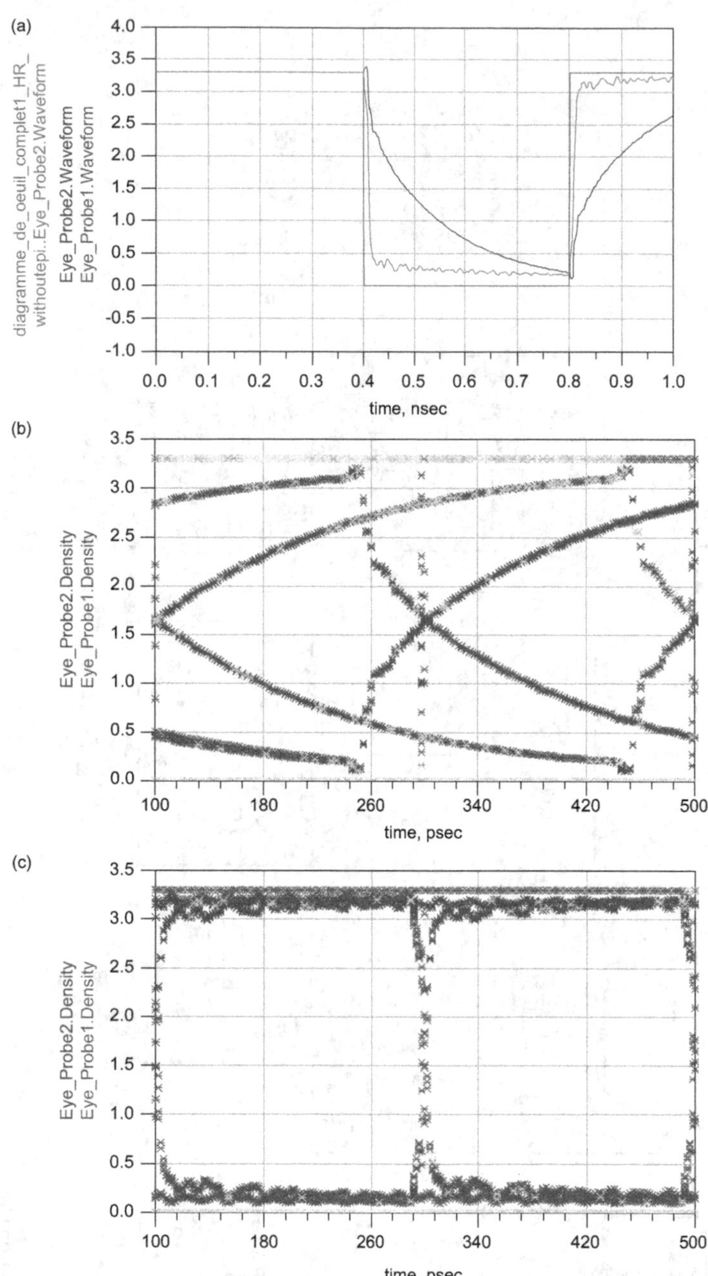

FIGURE 3.21
(a, d, and g) The input and output signals of the interconnection; (b, e, and h) the corresponding eye diagram for the low-resistive substrate; and (c, f, and i) the corresponding eye diagram for the high-resistive substrate.

(Continued)

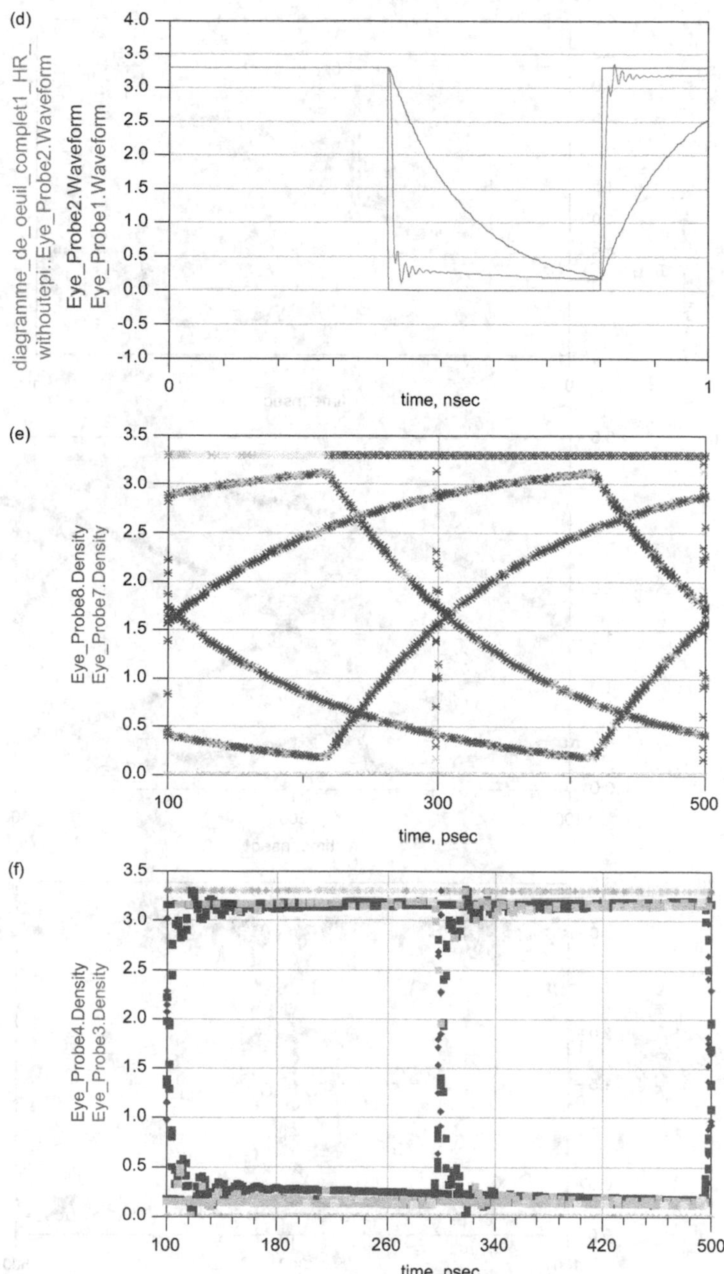

FIGURE 3.21 (CONTINUED)
(a, d, and g) The input and output signals of the interconnection; (b, e, and h) the corresponding eye diagram for the low-resistive substrate; and (c, f, and i) the corresponding eye diagram for the high-resistive substrate.

(Continued)

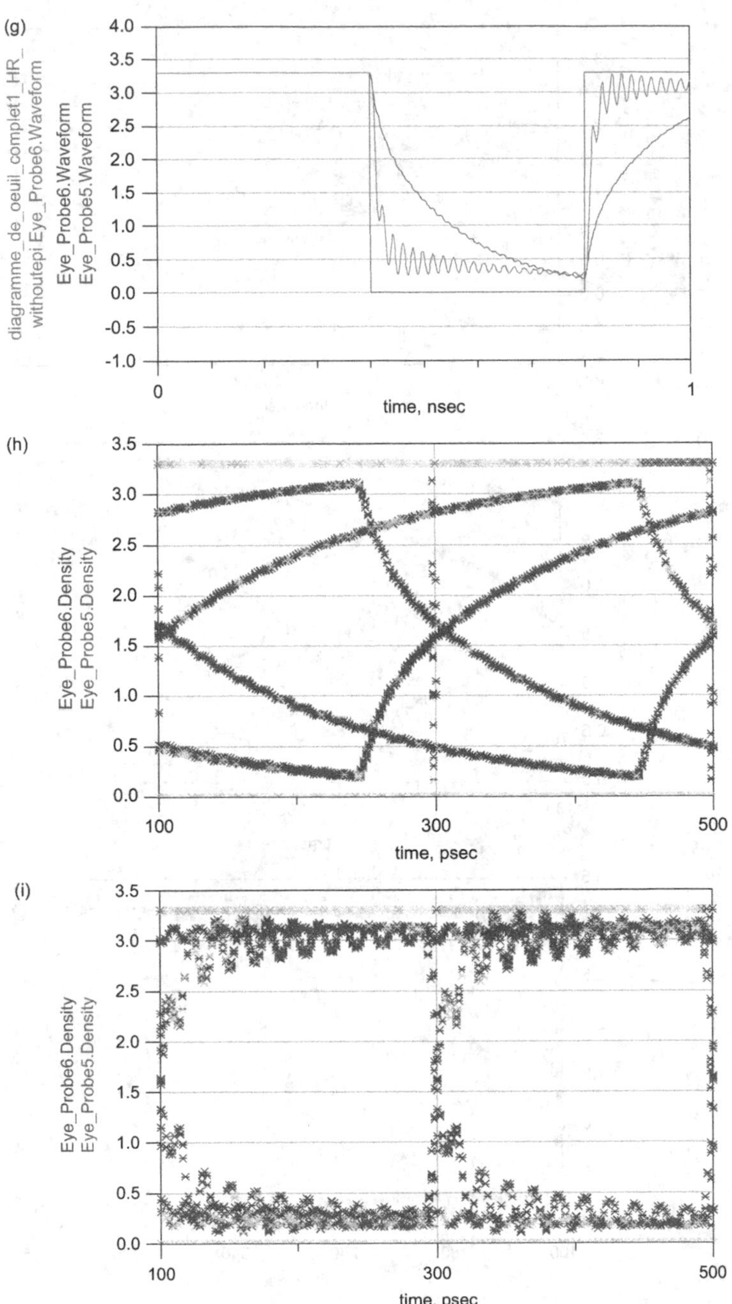

FIGURE 3.21 (CONTINUED)
(a, d, and g) The input and output signals of the interconnection; (b, e, and h) the corresponding eye diagram for the low-resistive substrate; and (c, f, and i) the corresponding eye diagram for the high-resistive substrate.

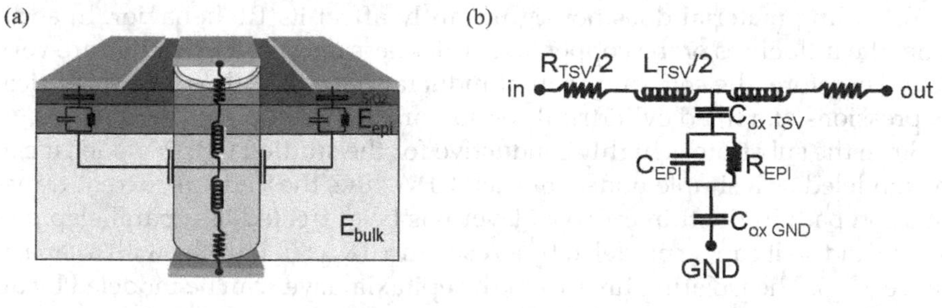

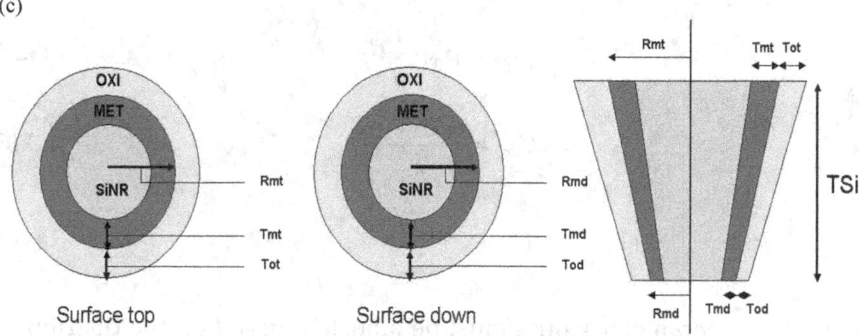

FIGURE 3.22
TSV schematic (a) cross-section view, (b) compact model, and (c) geometry parameters.

$$C_{\text{ox(TSV)}} = \frac{\varepsilon_0 \varepsilon_{\text{ox}} 2\pi H}{T_{\text{ot}} - T_{\text{od}}} \left[R_{\text{ot}} - R_{\text{od}} + \frac{R_{\text{ot}} \cdot T_{\text{od}} - R_{\text{od}} \cdot T_{\text{ot}}}{T_{\text{ot}} - T_{\text{od}}} \ln\left(\frac{T_{\text{od}}}{T_{\text{ot}}} \right) \right] \tag{3.12}$$

$$L_{\text{TSV}} = \frac{\mu_0 \mu_r H}{2\pi} \left[\sinh^{(-1)}\left(\frac{H}{R_a} \right) + \frac{R_a}{H} + 0.15 + \sqrt{\left(\frac{R_a}{H} \right)^2 + 1} \right] \tag{3.13}$$

$$M_{\text{TSV}} = \frac{\mu_0 \mu_r H}{2\pi} \left[\sin h^{-1}\left(\frac{H}{p} \right) + \frac{p}{H} - \sqrt{\left(\frac{p}{H} \right)^2 + 1} \right] \tag{3.14}$$

where R_{mt} and T_{mt} are the copper layer radius and thickness of the TSV top surface;

R_{ot} and T_{ot} are the oxide layer radius and thickness of the TSV top surface;

R_{md} and T_{md} are the copper layer radius and thickness of the TSV bottom surface;

R_{od} and T_{od} are the oxide layer radius and thickness of the TSV bottom surface;

p is the pitch between two nearby TSVs; and R_a is the TSV average radius

TSV filing material does not significantly affect its RF behavior. In addition, the influences of the copper layer thickness and the TSV radius are very low. Therefore, the self- and mutual-inductances can still use the analytical expressions of a filled cylindrical conductor.

Since the substrate is highly conductive for the studied CPWs model, it can be modeled as a simple node. For each CPW line, the vertical current transmission path in the thin epitaxial layer has been treated as a parallelepiped shape and so it can be modeled by a resistance (R_{epi}) in parallel with a capacitance (C_{epi}). The isolating lines from the epitaxial layer can be modeled by an oxide layer capacitances (C_{ox}) [59,60]:

$$R_{epi} = \rho_{epi}\frac{T_{epi}}{W \cdot L} \tag{3.15}$$

$$C_{epi} = \frac{\varepsilon_{epi}\varepsilon_0 W \cdot L}{T_{epi}} \tag{3.16}$$

$$C_{ox} = \frac{\varepsilon_{ox}\varepsilon_0 W \cdot L}{T_{ox}} \tag{3.17}$$

The electrical parameter values must be adjusted, based on the distributed elementary cell length, width, and height. For the TSV chain, the currents passe vertically through the epitaxial and the oxide layers and reach finally the BEOL level ground planes. The epitaxial and oxide layers electrical parameters (R_{epi}(TSV), C_{epi}(TSV) and C_{ox}(GND)) can be calculated by using the same expression as the coplanar waveguide. The substrate and TSVs couplings are considered with the geometrical capacitance C_{ox}(TSV). The coupling effects of TSVs are considered. The TSV chain electrical model of the BEOL ground lines is divided into serial blocks on the overall surface. These surface parts have a length of the TSVs contact pads length L_{pad}. The epitaxial layer parameters (R_{epi}(TSV), C_{epi}(TSV)) and the oxide capacitance (C_{ox}(GND)) which model the current propagation to the ground line through the TSV should then be calculated with $L = L_{pad}$. The width and the thickness are the same as the other line parts. The surface cells of ground lines included in the current paths can be simply modeled by RL networks; otherwise they are modeled as shown in the following Figure 3.23.

Figure 3.22 shows 2xTSV chain (a) medium-density TSV front view, (b) RLCG compact modeling of the medium-density TSV including the current path, and (c) decomposition of the BEOL ground lines into RLCG serial blocks depending on the considered surface (MI and MII models).

As an illustration, the above figures show how the two top BEOL ground lines are changed into serial blocks for a 2xTSV U-chain. The interline current paths should also be considered; they are mainly two kinds, MI and MII. The total wire length, L_{ground}, is equal to 2Lpad, plus the BRDL line length, L_{space}. L_{space} represents the gap between two contact pads. The BRDL line connects

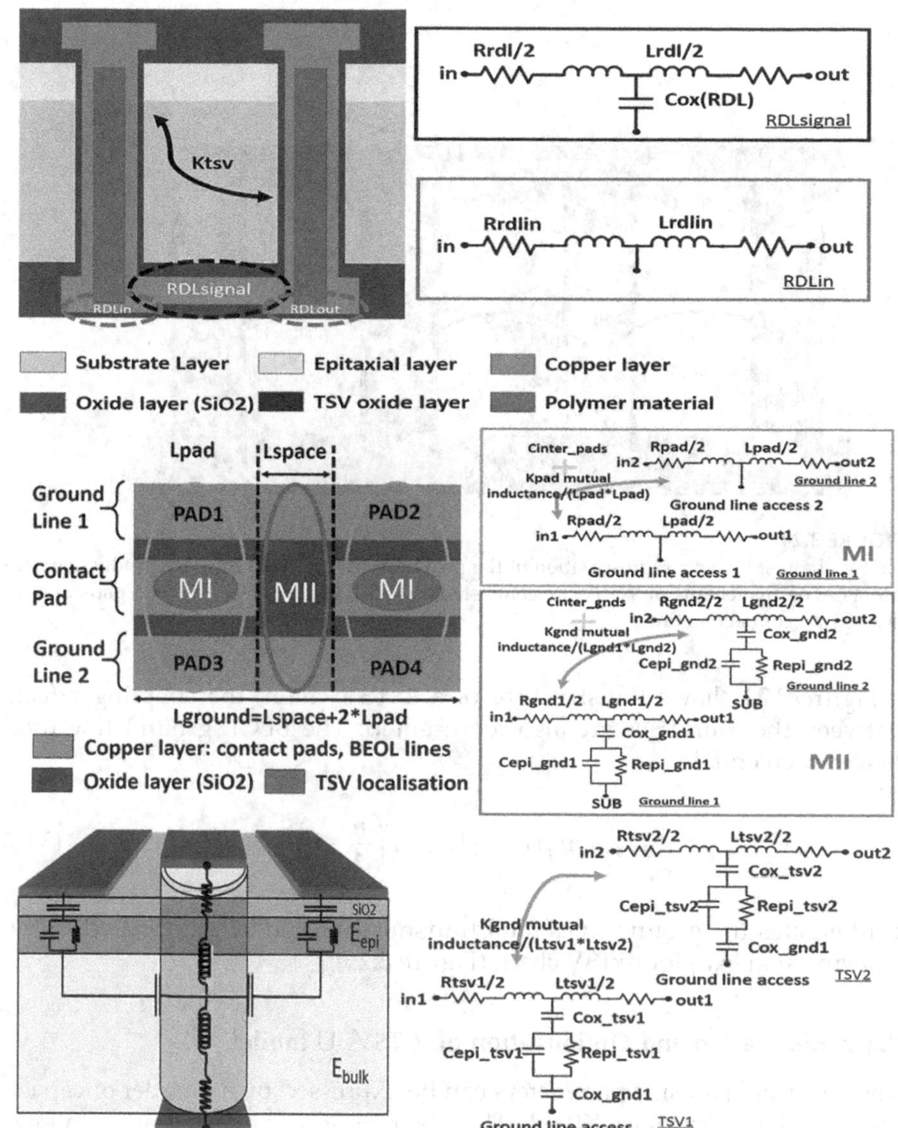

FIGURE 3.23
2xTSVs chain model geometry and compact model.

the two TSVs, the substrate coupling of which can be expressed by an oxide capacitance between the SUB node and BRDL, which is located in the contact pads gap (MII).

The 2xTSV has a form of "U". So, for nxTSVs chains, there will be $n/2$ "U-shaped" structures connected at the BEOL surface by contact pads and 60 μm access BEOL lines.

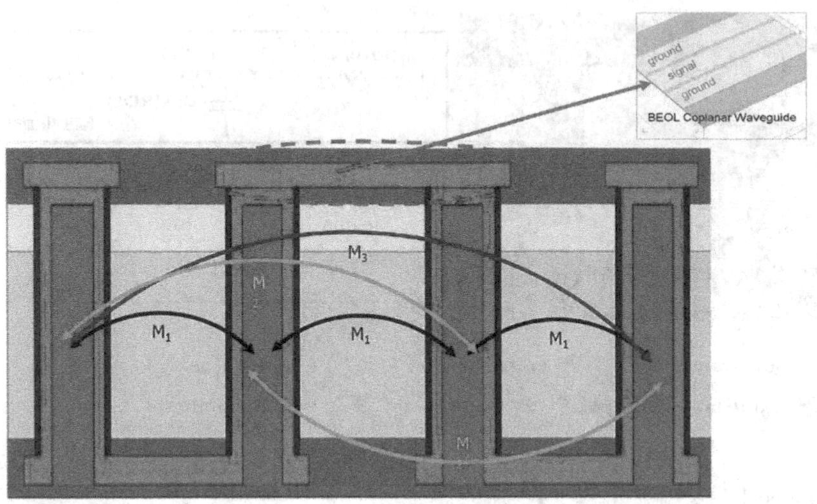

FIGURE 3.24
4xTSV chain side view: representation of the proximity effects between TSVs (inductive couplings). The two chains of 2xTSV are connected at the BEOL level with contact pads and an access interconnection line.

Figure 3.24 shows the structure of a 4xTSV chain; the coupling effects between the four TSVs are also represented. The BEOL ground line total length then equals:

$$L_{\text{BEOL}} = nL_{\text{pads}} + \frac{n}{2}L_{\text{space}} + \left(\frac{n}{2} - 1\right)L_{\text{access}} \quad (3.18)$$

It illustrates the ground lines total transmission line length (related to the transmission loss) for nxTSV chain (Figure 3.25).

3.4.2 Simulation and Optimization of 2 TSVs U model

The substrate parasitic parameters can be expressed by a parallel of capacitance and conductance [60,69]. Their expressions are shown as follows (Figure 3.26):

$$C_{(\text{Si\_sub})} = \frac{\pi \times \varepsilon_0 \varepsilon_{r,\text{si}}}{\cosh^{-1}\left(\frac{p_{\text{TSV}}}{d_{\text{TSV}}}\right)} \times h_{\text{TSV}}\,[F] \quad (3.19)$$

$$G_{\text{Si\_sub}} = \frac{\pi \times \sigma_{si}}{\cosh^{-1}\left(\frac{p_{\text{TSV}}}{d_{\text{TSV}}}\right)} \times h_{\text{TSV}}\,[S] \quad (3.20)$$

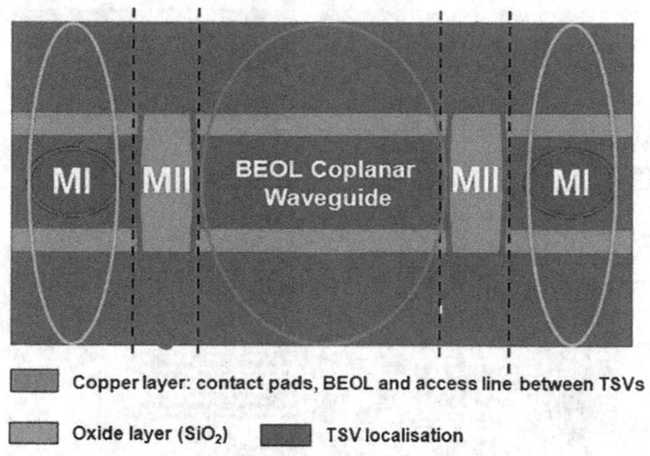

FIGURE 3.25
4xTSV chain top view: decomposition of the BEOL ground lines into serial blocks. Depending on the surface concerned, lines are modeled as a coplanar waveguide, MI or MII.

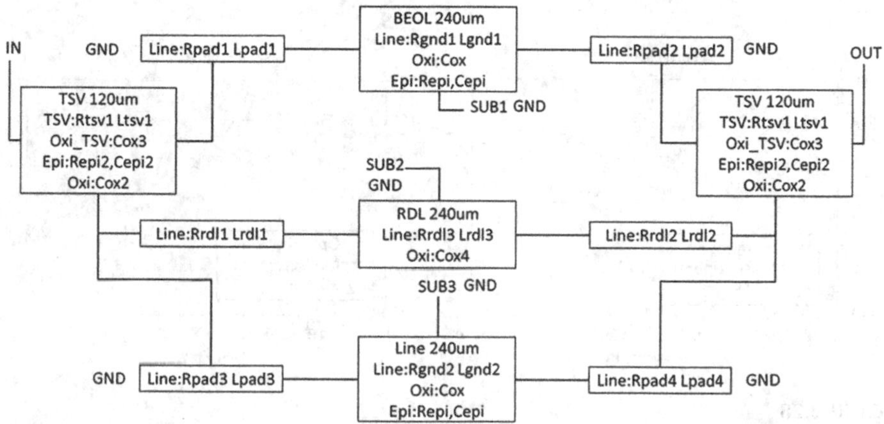

FIGURE 3.26
Input script description.

Here, when we extract the measurements, we must apply the de-embedding technique. Figure 3.27 shows the difference between with and without de-embedding.

As we can see from the two figures, the result with de-embedding is more close to our simulated circuits (Figure 3.28).

We present hereafter some comparisons between measurements and simulations for S-parameters (Figure 3.29).

As shown in Figure 3.30a,b, when $C_{ox3} = 2e^{-12}$ F, the curve is much closer to measurements. It corresponds to epsr_ox = 3.28 (we use 5.8 originally in

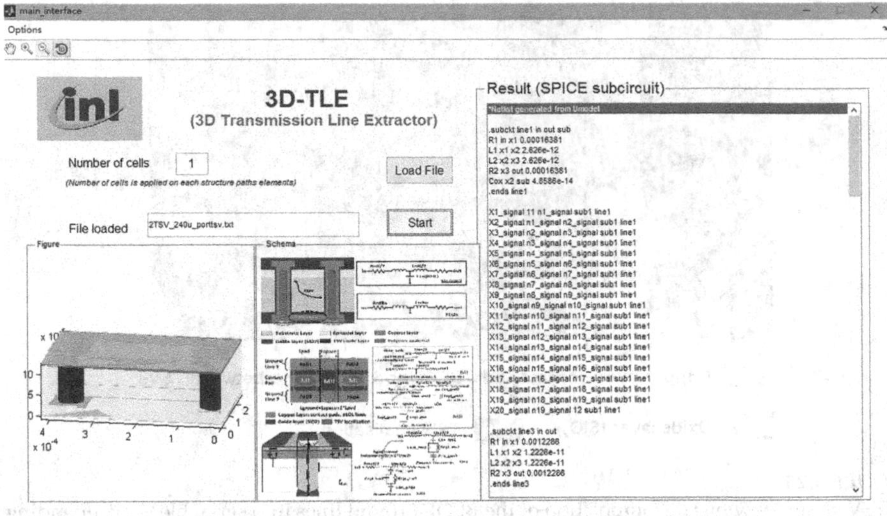

FIGURE 3.27
3D-TLE extractor setting.

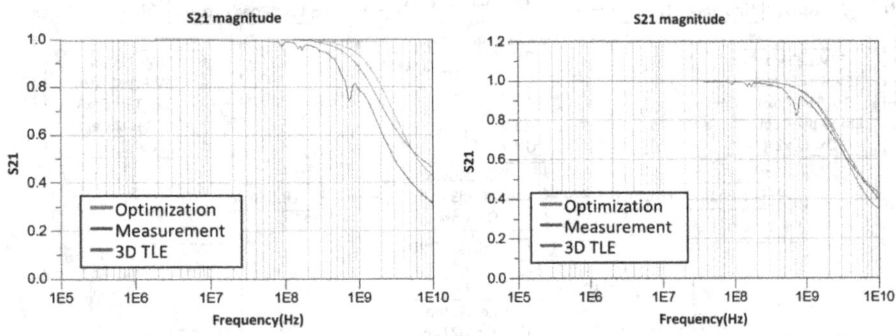

FIGURE 3.28
2xTSVs model *S*-parameters: (a) S21 magnitude without de-embedding and (b) S21 magnitude with de-embedding.

3D-TLE), and when $C_{ox2} = 7e^{-13}$ F, the curve is much closer to measurements. It corresponds to an epsr_ox = 10.98 (we use 5.8 originally, $C_{ox2} = 3.7e^{-13}$ originally); it has a very high influence on the final S-parameter approximation. R_{epi2} has a strong influence to the result. When $R_{epi2} = 38\ \Omega$, the result is close to the measurements. The 3D-TLE value is 50.5 Ω with a resistivity of $9.09e^{-2}$ Ω*m, now, with a $R_{epi2} = 38\ \Omega$, the resistivity changes to $6.84e^{-2}$ Ω*m. The MOS depletion effect does not have a big influence to the S21 parameters, but at very high frequency, it has a slightly influence. Here the C_{dep} vary from $0.5e^{-13}$ F to $8e^{-12}$ F.

As we can see, after 10 GHz, the substrate has a big influence to the S21 parameter. The best-fit R_{sub} value is about 100 Ω. At this resistance, the substrate

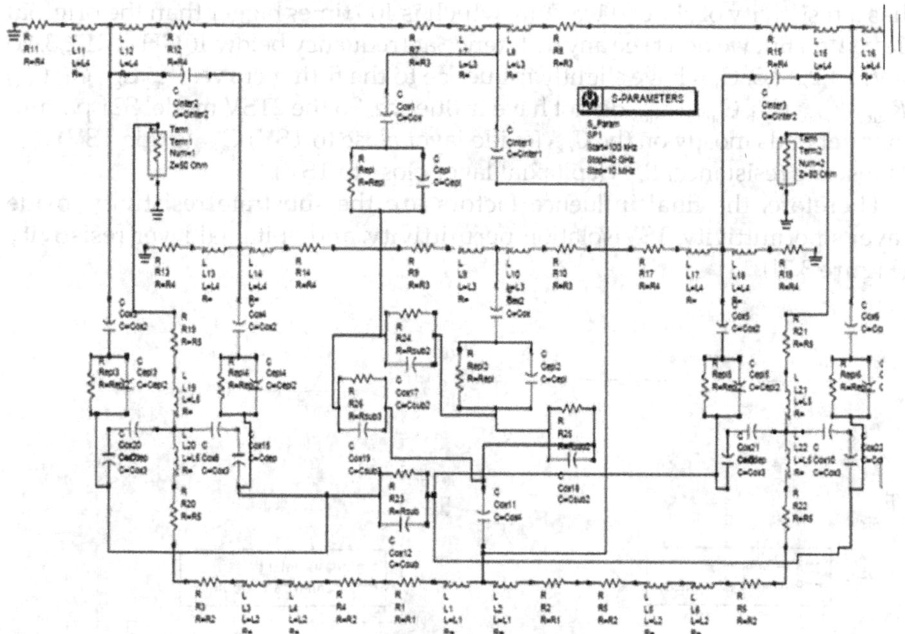

FIGURE 3.29
S-parameters study setting for 2xTSVs model (ADS schematics, 3D TLE).

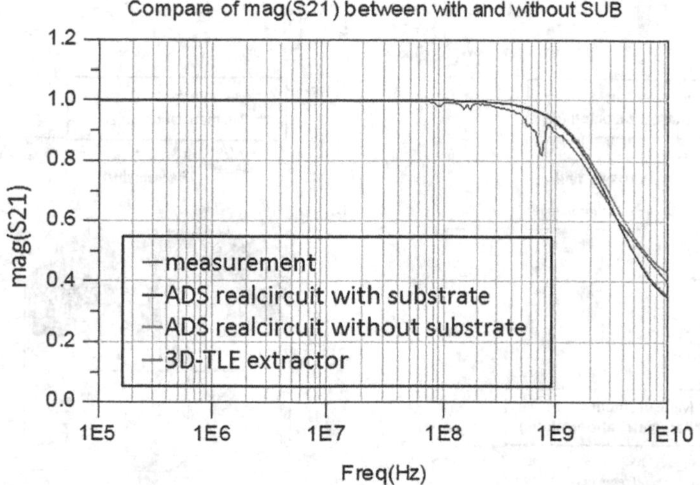

FIGURE 3.30
Comparison of S21 between 2TSV with and without SUB model, measurement (blue), optimized circuit result without SUB resistance (black), optimized circuit result with SUB resistance (red), and 3D-TLE result (pink).

has a resistivity of about 0.108 Ω*m, which is 100 times bigger than the original 1e^{-4} Ω*m. But we don't see any difference at frequency below 10 GHz. R1,2,3,4,5 and L1,2,3,4,5, C_{epi2} have slightly influence to the fitting curve. R_{epi} C_{epi} C_{ox} C_{ox4} R_{sub2} R_{sub3} C_{sub} C_{sub2} C_{sub3} do not have influence. So the 2TSV model S21 parameter depends mostly on the C_{ox2} (oxide layer close to TSV) C_{ox3} (oxide TSV) R_{sub} (substrate resistance) R_{epi2} (epitaxial layer close to TSV).

Therefore, the final influence factors are the substrate resistivity, oxide layer's permittivity, TSVisolation permittivity, and epitaxial layer resistivity (Figure 3.31).

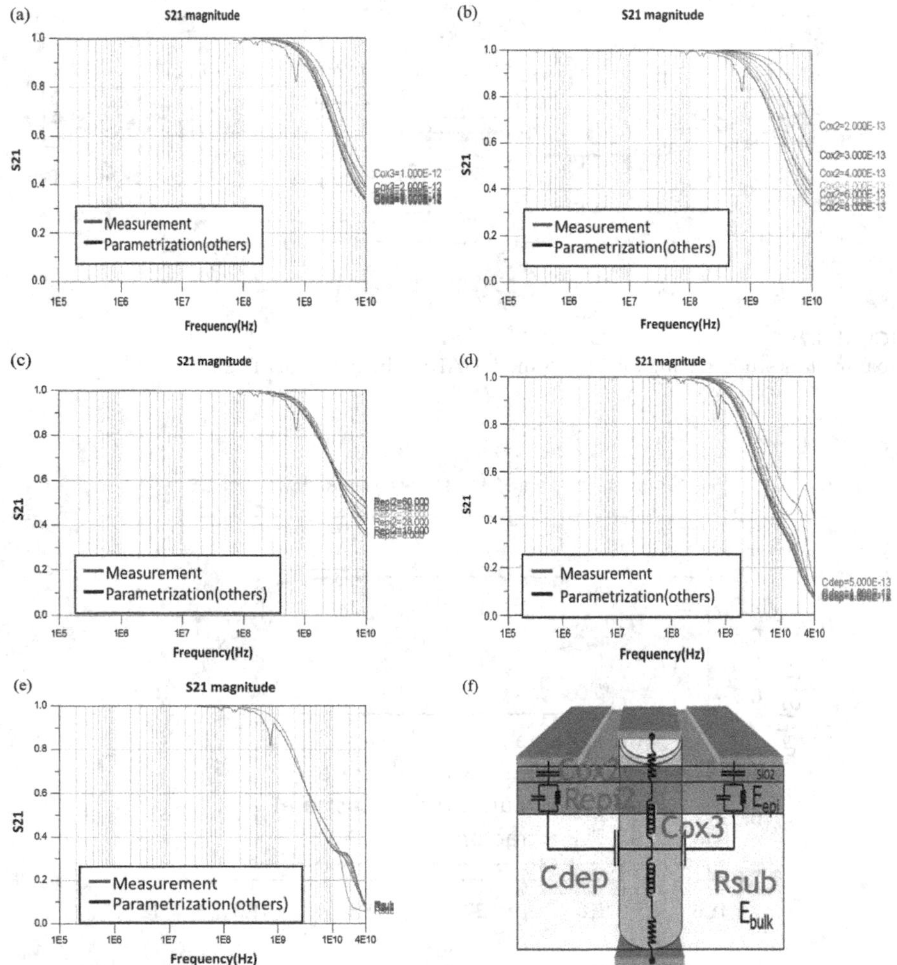

FIGURE 3.31
Parametrizations: (a) Cox3 (TSV oxide), (b) Cox2 (Oxide layer capacitance near TSV), (c) Repi2 (TSV epitaxial layer's resistance), (d) Cdep (TSV depletion capacitance), (e) Rsub (Substrate resistance), and (f) Parameters' signification.

After the optimization of the only four parameters, we can get the final S21 parameter.

With our model, the extracted parameters are given in Table 3.4.

With the optimized lumped parameters, we can find back the corresponding materials parameters, as shown in Table 3.5.

TABLE 3.4

Comparison between Original Parameters and Optimized Parameters

Name	Original Value	Optimized Value
R1	0.00016381	0.00016381
R2	0.0012286	0.0012286
R3	0.0016125	0.0016125
R4	0.012094	0.012094
R5	0.0031471	0.0031471
L1	$2.625e^{-12}$	$2.625e^{-12}$
L2	$1.2226e^{-11}$	$1.2226e^{-11}$
L3	$2.827e^{-12}$	$2.827e^{-12}$
L4	$1.3508e^{-11}$	$1.3508e^{-11}$
L5	$1.8915e^{-11}$	$1.8915e^{-11}$
R_{epi}	378.75	378.75
C_{epi}	$2.4851e^{-14}$	$2.4851e^{-14}$
C_{ox}	$4.9277e^{-14}$	$4.9277e^{-14}$
R_{epi2}	50.5	6.65094
C_{epi2}	$1.8638e^{-13}$	$1.8638e^{-13}$
C_{ox2}	$3.6958e^{-13}$	$0.73916e^{-12}$
C_{dep}	$0.66e^{-12}$	$1.32e^{-12}$
C_{ox3}	$1.9807e^{-12}$	$3.9614e^{-12}$
C_{ox4}	$4.8586e^{-14}$	$4.8586e^{-14}$
R_{sub}	1000	1994.4
R_{sub2}	1000	1000
R_{sub3}	1000	1000
C_{sub}	$1e^{-14}$	$1e^{-14}$
C_{sub2}	$1e^{-14}$	$1e^{-14}$
C_{sub3}	$1e^{-14}$	$1e^{-14}$

TABLE 3.5

Restored Layers' Material Parameters

	Marked Value	Optimized Value
Relative Permittivity TSV oxide layer	5.8	6.5001
Relative Permittivity oxide layer	5.8	11.6001
Resistivity substrate (Ω*m)	$1e^{-4}$	0.2154
Resistivity epitaxial (Ω*m)	$9.09e^{-2}$	$1.2e^{-2}$

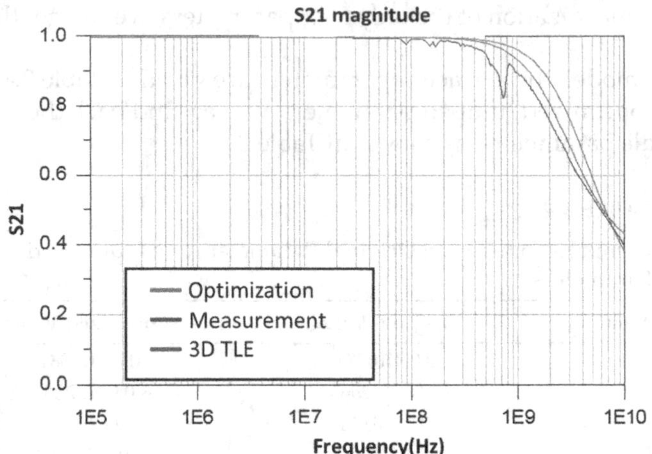

FIGURE 3.32
Optimized *S*-parameters (blue: measurement, red: optimized, pink: 3D-TLE).

As we can see, the TSV oxide relative permittivity has changed from 5.8 to 6.5, that is close; there may be some measurement's error—see the resistivity of the epitaxial layer. The oxide layer relative permittivity has doubled to 11.6, and the substrate is no more non-resistive; its resistivity changes to 0.2154 Ω·m.

In order to test more our 3D-TLE extractor, a finite-element simulation (COMSOL Multiphysics [43]) is performed. The simulation results are shown in Figure 3.32. They have a really good fit between 3D-TLE extractor, COMSOL simulation, and measurement.

3.4.3 TSV Matrix Model

3.4.3.1 Study of Signal Integrity in TSV Matrices

Since the 3D interconnect electrical models have been validated in low- and medium-frequency ranges, timing analyses are now proposed, using Eye Diagrams to examine the signal integrity in the case of a 3 × 3 medium-density TSVs regular matrix. The matrix is modeled with W1 configuration TSVs. The pitch between each TSV is 120 μm. The signal integrity is studied at low and medium frequencies, so we make the assumption that all the TSV oxide capacitances join at a single node representing the conductive substrate. This modeling consideration is illustrated in Figure 3.33.a for a 2 × 2 TSVs matrix. All the capacitances between neighbor TSVs are taken into account with inter-pillar parasitic capacitance formula. Concerning magnetic field coupling, only the TSV proximity effects (mutual inductive couplings) are considered in these analyses.

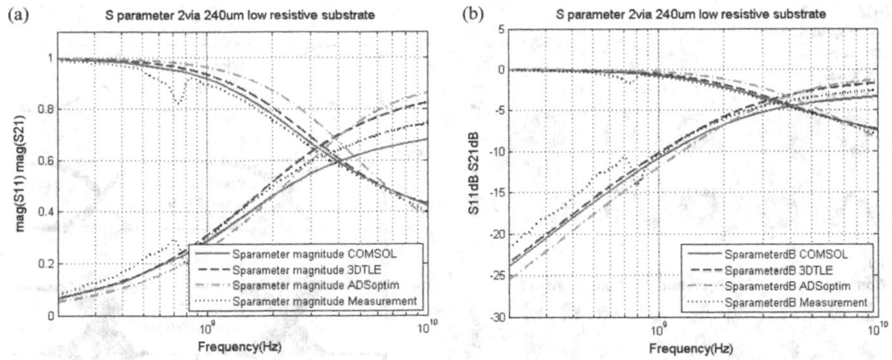

FIGURE 3.33
3D-TLE verification by using FEM (COMSOL), ADS and measurements: (a) Magnitude of *S*-parameters and (b) dB of *S*-parameters.

The Eye Diagrams are produced by assuming in the simulation a symmetrical test bench, with 50 Ω output buffer equivalent resistances connected to each TSV input. The TSVs are terminated with 4fF load capacitances. The matrix SPICE netlist is generated with our 3D-TLE extractor and exported to Cadence Virtuoso Spectre [72] for circuit timing analyses. Depending on the studied configuration, "0" logic signals or pseudo-random bit sequences, having a magnitude of 1 V and a rise time of 500 ps for data rates of 500 Mb/s, are transmitted to the TSVs.

Three configurations are investigated. For the first one (A), all the TSVs are grounded except the central TSV. The second configuration (B) is a shielded arrangement for which data are transmitted through the matrix's inner TSVs; the outer ones are grounded. In the third (C) configuration, data are transmitted to through all the TSVs (Figure 3.34).

3.4.4 Electrical Modeling Result of 3D IC

3.4.4.1 Case I: Electrical Modeling of 3xRLC Segments Model

The model can be described through a SPICE-like script (Figure 3.35).

It concerns three-segment *RLC* circuits. The electrical circuit is shown in Figure 3.36.

The segments' resistance is 3.5 Ω, inductance is $1.2e^{-3}$ H, the capacitance is $7.3e^{-6}$ F, and the added capacitance at the probe side is $10e^{-6}$ F. The voltage source is a sine signal, with amplitude of 3 V, and working at frequency 200 Hz. Here, we use the modified nodal analysis method (MNA) and generalized conjugate residual (GCR) solver and also sparse matrix technique to solve and accelerate the calculation of large scale differential matrix. In order to verify the method's robustness, the same simulation is done with ADS (advanced design system by Keysight Technologies [64]). The signal is

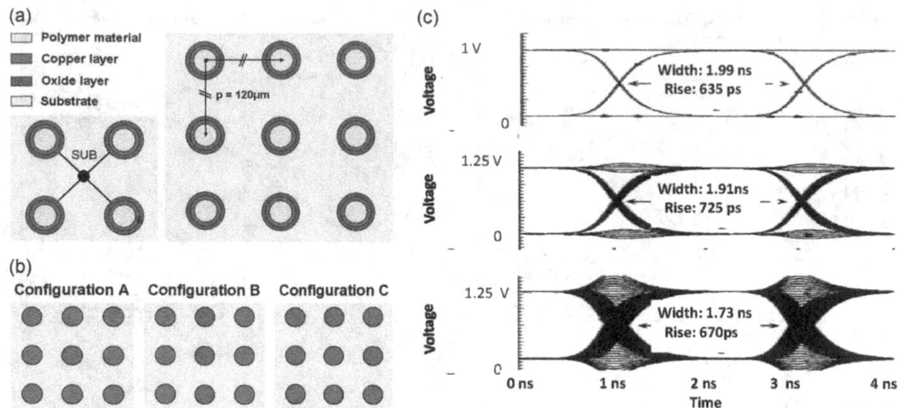

FIGURE 3.34
(a) Medium-density TSVs regular matrices. Bottom left: all the TSVs' oxide capacitances join at a single node modeling the substrate (illustration for a 2 × 2 matrix). Right: 3 × 3 medium-density TSVs matrix considered for the timing domain analyses. (b) 3 × 3 medium-density TSV matrix configurations for generating Eye Diagrams from simulations. Color legend: gray for grounded TSV, red for signal. Configuration A: data are only transmitted through the matrix central TSV. Configuration B: shielded configuration. Configuration C: data are sent to all the TSVs. (c) Generating Eye Diagrams from simulation based on the three studied configurations. Top: A configuration. Middle: B configuration. Bottom: C configuration.

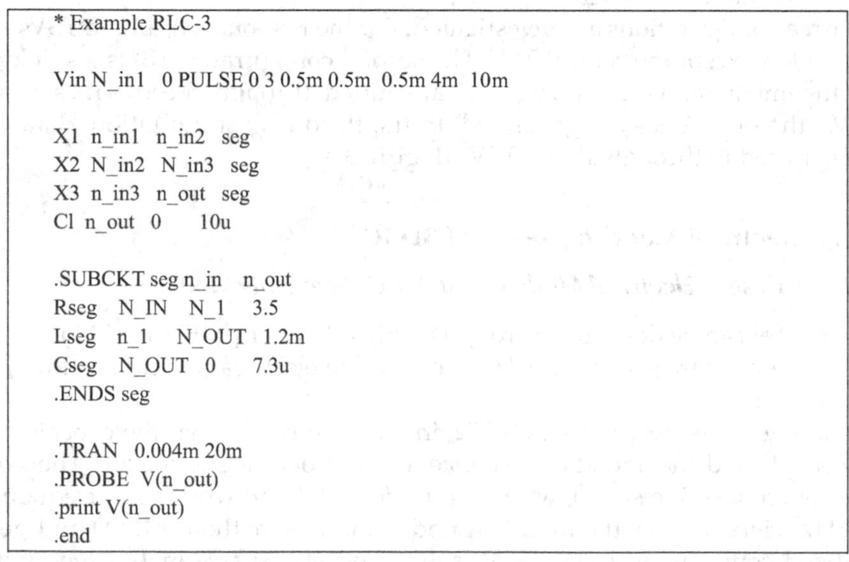

```
* Example RLC-3

Vin N_in1  0 PULSE 0 3 0.5m 0.5m  0.5m 4m  10m

X1  n_in1  n_in2  seg
X2  N_in2  N_in3  seg
X3  n_in3  n_out  seg
Cl  n_out  0      10u

.SUBCKT seg n_in   n_out
Rseg   N_IN   N_1    3.5
Lseg   n_1    N_OUT  1.2m
Cseg   N_OUT  0      7.3u
.ENDS seg

.TRAN   0.004m 20m
.PROBE  V(n_out)
.print V(n_out)
.end
```

FIGURE 3.35
3xRLC series model SPICE script.

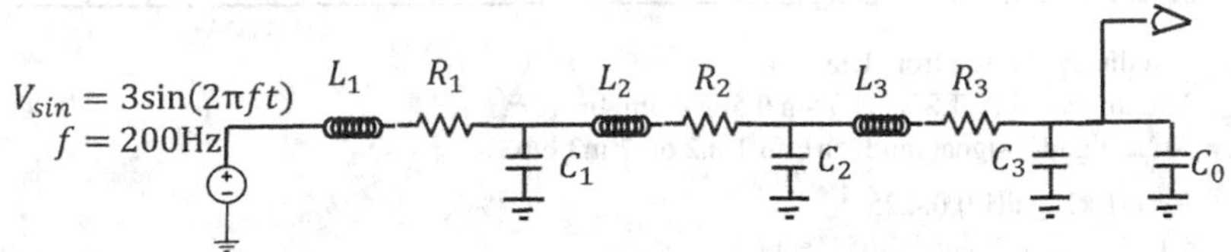

FIGURE 3.36
Compact 3 RLC segments model.

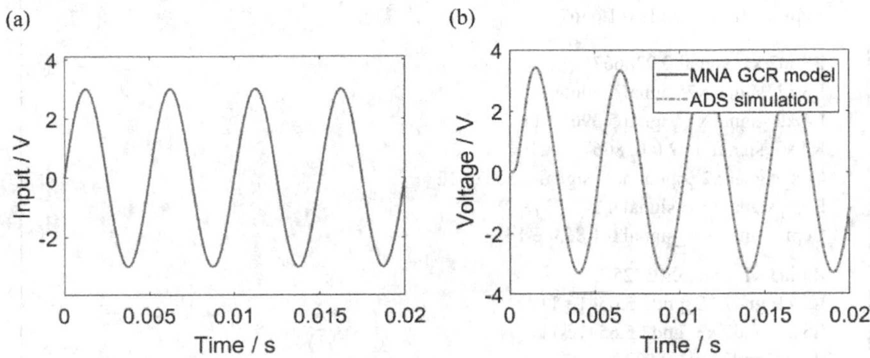

FIGURE 3.37
(a) Input voltage source. (b) Output comparison between MNA and ADS. (c) Relative error between MNA and ADS for 3 RLC segments model.

shown in Figure 3.37a and the output signal comparison with ADS is shown in Figure 3.37b. The MNA+GCR method has a very good accuracy in comparing with the ADS; the maximum error is about 4%.

3.4.4.2 Case II: CPW (Coplanar Wave Model)

As the same, but slightly different, we define the model's geometry and material parameters in one txt file, and then by using INL's 3D-TLE (3D Transmission line extractor), generate automatically a SPICE-compatible sp file. This file is shown in Figure 3.38.

The model describes a coplanar wave model. The electrical circuit is recall in Figure 3.39.

We define two kinds of lines: one signal line and one ground line. Each line is treated as a T model. We define a C_{inter} between any two lines and each line is isolated from the substrate by an oxide layer. The voltage source is a pulse signal, with amplitude of 1 V, rise and down time of 0.05 ns, duty cycle of 60%, and working frequency 2 GHz. Here, we use still the MNA and GCR solver. In order to verify the method's robustness, the same simulation is done with ADS. The input signal is shown in Figure 3.40a and the output

```
*Netlist generated from Umodel
Vin in_real  0 PULSE 0 3 0.5m 0.5m  0.5m 4m 10m
.subckt gnd1_signal_gnd2 in1 out1 in2 out2 in3 out3
R1 in1 x1_gnd1 0.03225
L1 x1_gnd1 x2_gnd1 5.6541e-11
L2 x2_gnd1 x3_gnd1 5.6541e-11
R2 x3_gnd1 out1 0.03225
Cox_gnd1 x2_gnd1 nox_gnd1 1.3335e-12
Repi_gnd1 nox_gnd1 0 312.5
Cepi_gnd1 nox_gnd1 0 1.6567e-14

R3 in2 x1_signal 0.028667
L3 x1_signal x2_signal 5.3965e-11
L4 x2_signal x3_signal 5.3965e-11
R4 x3_signal out2 0.028667
Cox_signal x2_signal nox_signal 1.5002e-12
Repi_signal nox_signal 0 277.7778
Cepi_signal nox_signal 0 1.8638e-14

R5 in3 x1_gnd2 0.03225
L5 x1_gnd2 x2_gnd2 5.6541e-11
L6 x2_gnd2 x3_gnd2 5.6541e-11
R6 x3_gnd2 out3 0.03225
Cox_gnd2 x2_gnd2 nox_gnd2 1.3335e-12
Repi_gnd2 nox_gnd2 0 312.5
Cepi_gnd2 nox_gnd2 0 1.6567e-14

K1 L1 L3 0.42787
K2 L2 L4 0.42787
Cinter1 x2_signal x2_gnd1 6.7837e-15

K3 L3 L5 0.42787
K4 L4 L6 0.42787
Cinter2 x2_gnd2 x2_signal 6.7837e-15

K5 L1 L5 0.30072
K6 L2 L6 0.30072
.ends gnd1_signal_gnd2

X1_gnd1_signal_gnd2 0 0 1 2 0 0 gnd1_signal_gnd2
C1 2 0 100e-15
R1 in_real 1 50
.TRAN  0.004m 20m
.PROBE  V(2)
.print V(2)
.end
.end
```

FIGURE 3.38
CPW model SPICE script.

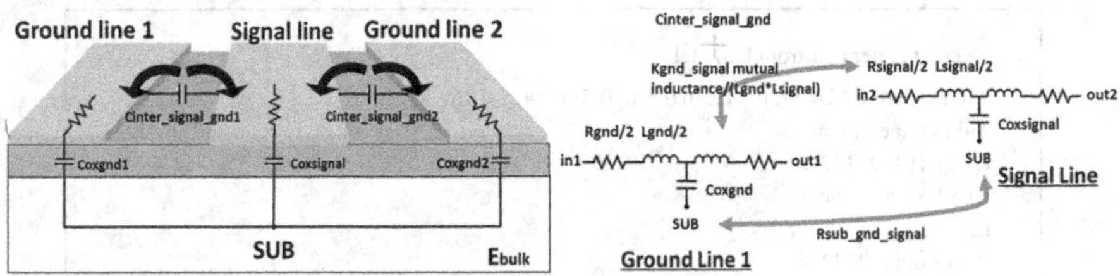

FIGURE 3.39
CPW geometry and compact model.

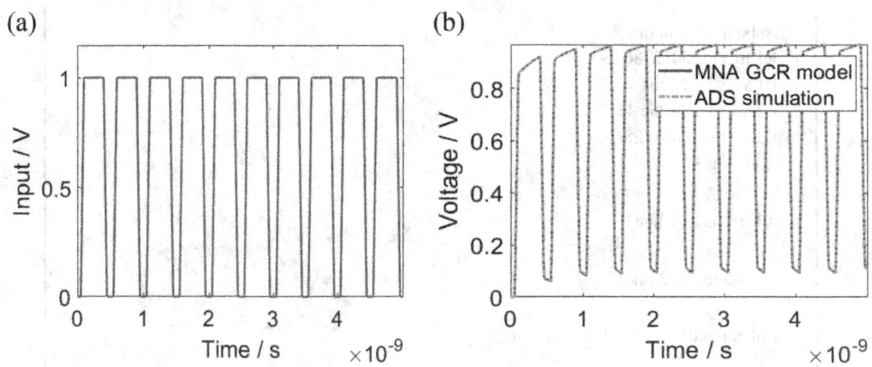

FIGURE 3.40
CPW model: (a) input voltage source and (b) output comparison between MNA and ADS.

signal comparison with ADS is shown in Figure 3.39b. Our MNA+GCR method has a very good accuracy, compared with the ADS's one; the maximum error is about 3.5%.

3.4.4.3 Case III: Two TSVs U model [62,73–81]

We define, here, the model's geometry and material parameters in one text file, and then by using our (cf. Chapter 2) 3D-TLE (3D Transmission line extractor), generate automatically a SPICE-compatible sp file. This file is shown in Figure 3.41.

We define two kinds of lines: signal line and ground line, and each line is treated as a T model. The exact model shown in the upper voltage source is a pulse signal, with an amplitude of 1 V, rise and down time of 0.05 ns, duty cycle of 60%, and working frequency 2 GHz. Here we also use the MNA and GCR solver. In order to verify the method's robustness, the same simulation is done in ADS. The signal is shown in Figure 3.37a and the output signal comparison with ADS is shown in Figure 3.37b. The MNA+GCR method has a very good accuracy as compared with the ADS; the maximum error is about 1.6%.

```
*Netlist generated from Umodel
Vin in_real  0 PULSE 0 3 0.5m 0.5m  0.5m   4m  10m
.subckt line1 in out
R1 in x1 0.0032762
L1 x1 x2 5.2519e-11
L2 x2 x3 5.2519e-11
R2 x3 out 0.0032762
Cox x2 0 9.7173e-13
.ends line1

X1_signal 11 12 line1

.subckt line3 in out
R1 in x1 0.0012286
L1 x1 x2 1.2226e-11
L2 x2 x3 1.2226e-11
R2 x3 out 0.0012286
.ends line3

X1_rdlin 1 11 line3

X1_rdlout 12 2 line3

.subckt gnd1_gnd2 in1 out1 in2 out2

R1 in1 x1_gnd1 0.03225
L1 x1_gnd1 x2_gnd1 5.6541e-11
L2 x2_gnd1 x3_gnd1 5.6541e-11
R2 x3_gnd1 out1 0.03225
Cox_gnd1 x2_gnd1 nox_gnd1 9.8554e-13
Repi_gnd1 nox_gnd1 0 18.9375
Cepi_gnd1 nox_gnd1 0 4.9702e-13

R3 in2 x1_gnd2 0.03225
L3 x1_gnd2 x2_gnd2 5.6541e-11
L4 x2_gnd2 x3_gnd2 5.6541e-11
R4 x3_gnd2 out2 0.03225
Cox_gnd2 x2_gnd2 nox_gnd2 9.8554e-13
Repi_gnd2 nox_gnd2 0 18.9375
Cepi_gnd2 nox_gnd2 0 4.9702e-13

K1 L1 L3 0.30072
K2 L2 L4 0.30072
Cinter1 x2_gnd2 x2_gnd1 1.1493e-15

.ends gnd1_gnd2
```

FIGURE 3.41
2xTSVs chain U model.

(Continued)

```
X1_gnd1_gnd2 3 4 5 6 gnd1_gnd2

.subckt pad1_pad3 in1 out1 accesstsv1 in2 out2 accesstsv2

R1 in1 x1_pad1 0.012094
L1 x1_pad1 accesstsv1 1.3508e-11
L2 accesstsv1 x2_pad1 1.3508e-11
R2 x2_pad1 out1 0.012094

R3 in2 x1_pad3 0.012094
L3 x1_pad3 accesstsv2 1.3508e-11
L4 accesstsv2 x2_pad3 1.3508e-11
R4 x2_pad3 out2 0.012094

K1 L1 L3 0.74376
K2 L2 L4 0.74376
Cinter1 accesstsv2 accesstsv1 4.3097e-16

.ends pad1_pad3

X1_pad1_pad3 0 3 101 0 5 102 pad1_pad3


.subckt pad2_pad4 in1 out1 accesstsv1 in2 out2 accesstsv2

R1 in1 x1_pad2 0.012094
L1 x1_pad2 accesstsv1 1.3508e-11
L2 accesstsv1 x2_pad2 1.3508e-11
R2 x2_pad2 out1 0.012094

R3 in2 x1_pad4 0.012094
L3 x1_pad4 accesstsv2 1.3508e-11
L4 accesstsv2 x2_pad4 1.3508e-11
R4 x2_pad4 out2 0.012094

K1 L1 L3 0.74376
K2 L2 L4 0.74376
Cinter1 accesstsv2 accesstsv1 4.3097e-16

.ends pad2_pad4

X1_pad2_pad4 4 0 201 6 0 202 pad2_pad4
```

FIGURE 3.41 (CONTINUED)
2xTSVs chain U model.

(Continued)

```
.subckt tsv1_tsv2 in1 out1 side1 side2 in2 out2 side3 side4

R1 in1 x1_tsv1 0.0031471
L1 x1_tsv1 x2_tsv1 1.8915e-11
L2 x2_tsv1 x3_tsv1 1.8915e-11
R2 x3_tsv1 out1 0.0031471
Coxtsv1_tsv1 x2_tsv1 p1_tsv1 1.9807e-12
Coxtsv2_tsv1 x2_tsv1 p2_tsv1 1.9807e-12
Repi1_tsv1 p1_tsv1 epi1_tsv1 50.5
Cepi1_tsv1 p1_tsv1 epi1_tsv1 1.8638e-13
Repi2_tsv1 p2_tsv1 epi2_tsv1 50.5
Cepi2_tsv1 p2_tsv1 epi2_tsv1 1.8638e-13
Coxgnd1_tsv1 epi1_tsv1 side1 3.6958e-13
Coxgnd2_tsv1 epi2_tsv1 side2 3.6958e-13

R3 in2 x1_tsv2 0.0031471
L3 x1_tsv2 x2_tsv2 1.8915e-11
L4 x2_tsv2 x3_tsv2 1.8915e-11
R4 x3_tsv2 out2 0.0031471
Coxtsv1_tsv2 x2_tsv2 p1_tsv2 1.9807e-12
Coxtsv2_tsv2 x2_tsv2 p2_tsv2 1.9807e-12
Repi1_tsv2 p1_tsv2 epi1_tsv2 50.5
Cepi1_tsv2 p1_tsv2 epi1_tsv2 1.8638e-13
Repi2_tsv2 p2_tsv2 epi2_tsv2 50.5
Cepi2_tsv2 p2_tsv2 epi2_tsv2 1.8638e-13
Coxgnd1_tsv2 epi1_tsv2 side3 3.6958e-13
Coxgnd2_tsv2 epi2_tsv2 side4 3.6958e-13

K1 L1 L3 0.11406
K2 L2 L4 0.11406

.ends tsv1_tsv2

X1_tsv1_tsv2 in 1 101 102 2 out 201 202 tsv1_tsv2
R1_tsv1 in_real in 50
C1_tsv2 out 0 100e-15
.TRAN  0.004m 20m
.PROBE  V(out)
.print V(out)
.end
.end
```

FIGURE 3.41 (CONTINUED)
2xTSVs chain U model.

In Figure 3.42, we can find out that the MNA+GCR method has the same accuracy with ADS simulator, and it can be integrated with 3D-TLE extractor without any difficulties. The general design process is shown in Figure 3.43.

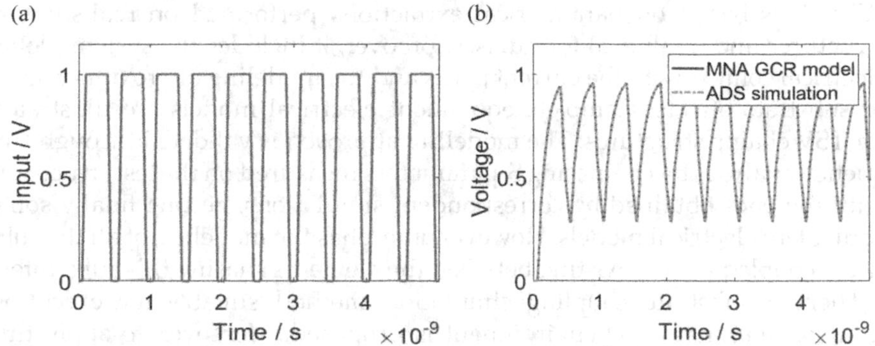

FIGURE 3.42
(a) Input voltage source. (b) Output comparison between MNA and ADS. (c) Relative error between MNA and ADS for 2xTSVs chain model.

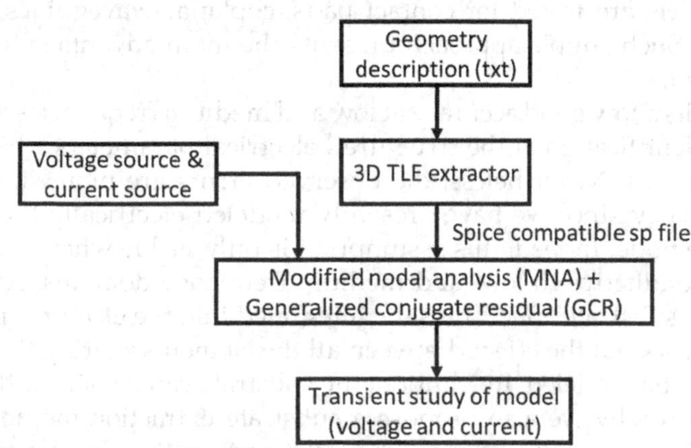

FIGURE 3.43
Our workflow for 3D device design.

This part gives a simple thermal modeling solution for 3D IC. This study method is viable for one CPW model; it can also be used for 20, 100, and more CPWs cases. A difficulty lies on that we could not consider a great number of CPWs, as simple sum of CPWs; we do consider non-linear models, which needs more physical studies.

3.5 Conclusion

We propose some compact and physical modeling approach of the interconnect part: it is applied essentially to redistribution lines and medium-density

TSVs. It is based on parametrical extractions performed on realistic test structures and analytical formulas. Moreover, it includes the system global electrical context with the current paths and the modeling of proximity and/or substrate effects. Complete equivalent electrical models are illustrated for TSV chains structures. The modeling approach is validated through frequency analyses by comparing S-parameters measured on the test structures with the ones obtained by correspondent simulations, getting finally some equivalent electrical models. However, it implies the modeling of all the substrate coupled effects existing between the elements sharing this substrate.

Then, a substrate coupling simulation method suitable for execution in a conventional CAD environment is proposed. Moreover, to apply this approach to a realistic mixed signal design, several challenges have to be grasped. A key one concerns the way to obtain, for instance, the transient consumption current of an entire large digital part. Some CAD tools offer the possibility to simulate these currents.

Our models are tested for contact pads, coplanar waveguides, and TSV structures. Such simple approach presents the main advantage to be easily automatized.

The results show good accuracy at low and medium frequencies and enable the clear identification of the structure's electrical parameters which impact insertion losses. Nevertheless, the observed errors are more significant at high frequency since we have presently modeled electrically the substrate as a simple node. Indeed, this assumption is only viable when the substrate is highly conductive in low- and medium-frequency domains. At high frequency, the substrate effects must be integrated into the electrical model as a *RLCG* network but the effects between all the elements sharing the substrate have also to be modeled. In addition, the substrate can be non-uniform. This is the reason why we also propose a substrate extraction method, relying on the TLM over multi-layered substrates and on the Green functions, to model these effects. The method efficiency has been tested for contact pads and coplanar waveguides structures. Such simple approach presents the main advantage to be easily automatized; additionally it can be improved by including skin effect and eddy current. Ongoing works will consist in extending the modeling approach to other 3D interconnect types, such as copper pillars for which a model has been proposed in [2] without inductive coupling consideration, and to different types of 3D interconnect matrices. The substrate extractor will be incorporated in cases of chains and matrices of TSVs to electrically model, with increased accuracy, a large panel of 3D structures and critical paths and to investigate signal integrity.

4

Electrothermal Modeling of Substrates

4.1 Introduction

Substrate noise coupling in integrated circuits (ICs) becomes a significant consideration in the circuit design. Nowadays, micro (nano) technology and the development of semiconductor technology enable designers to integrate multiple systems into a single chip, not only in 2D (planar) but also in 3D (in the bulk). This design technology reduces cost, while improving performance, and makes the possibility of a system to be on chip [2,3,4].

3D Si integration seems to be the right way to go and compete with Moore's law (more than Moore versus more Moore). However, it is still a long way to go. In 2010, the question was why 3D. Today, the questions are when and how is 3D used. The 3D chip stacking is well known to overcome conventional 2D IC issues, using in-depth contacts or some through silicon via for signal transmission.

The kind of silicon substrate in which 3D interconnections are processed is an important parameter and has a strong impact on 3D interconnection electrical behavior [19]. Electrical modeling methods are dependent on the substrate type of the application. Two different kinds of p silicon substrates are typically used in CMOS/BiCMOS processes: uniform lightly doped substrate (>1 Ω·cm) and heavily doped substrate (<1 Ω·cm) with lightly thinned doped (>1 Ω·cm) epitaxial layer. Other semiconductor substrates are used in micro/nanotechnologies and could embed 3D interconnections (silicon on insulator technologies, microelectromechanical systems technologies, sensor technologies, etc.).

Indeed, we attempt to propose an analysis of dynamic electrothermal phenomenon studies in 3D layered circuits.

First of all, we solve Poisson's equation, using Green kernels [82,83]. Here, we show that this mathematical method is an analog to the transmission line matrix (TLM) method, based on reflection/transmission of voltage and current through the layers, when adopting it to embedded contacts of any shape. Then, we work on the resolution of the unsteady 3D heat equation; during this, we use the algorithms developed for the electrical modeling, particularly when using the spreading impedance concept. Finally, we show that these electrical and thermal spreading impedances can be considered entangled. This approach drives us to start an attempt to model the more general problem of noise in 3D ICs (tri-dimensional integrated circuits), using again, a transfer impedance paradigm.

4.2 Substrate Modeling Approach

In this part, Green functions are applied to homogeneous stacked layers for substrate model extraction, as opposed to numerical methods; the resolution speed of the former method is much faster. Some basic recalls and concepts are first introduced. The use of discrete cosine transforms (DCTs), applying twice a fast Fourier transform (FFT) in this model, will accelerate and compute the speed. Then, an improved model, which can be applied on substrate with in-depth contacts (or via), is shown; it can treat the case of contacts lying in different layers. For verification, a finite-element method (FEM) is compared [43].

4.2.1 Green Kernels/TLM

The high-density integration and high system frequency make substrate noise coupling one of the most significant considerations in the design due to its great impact on the performance of ICs. The aim of this 3D substrate analysis is, first of all, to efficiently extract the Z impedance parameters between any contacts (electrode and via) that are located onto or into the silicon substrate. An efficient impedance extraction tool, for any two contacts, could help the designer to accelerate the design and optimize the final layout. Like in planar technologies, 3D interconnects can be built as a resistance, inductance, capacitance, and conductance (RLCG) equivalent electrical model, with a Π or a T network. Generally, a simple compact model is constructed by modeling the substrate as a simple node. However, this assumption is only viable when substrate is highly conductive, in low- and medium-frequency domains, and is not suitable for the multilayer substrates. That is the reason why, in this work, we propose a substrate extraction method relying on TLM method over multilayered substrate and/or Green functions to model these effects, in the bulk [82–91].

4.2.2 Substrate Analytical Modeling

The most well-known pioneering paper came from ref. [82]: a layered 3D substrate, with two surface contacts.

We present hereafter a 3D substrate study with top or embedded contacts. The originality as we consider and apply here is to do all the calculations in the reciprocal domain, from the very beginning of the calculations to the end, to extract voltage or temperature.

In general, the Z-parameters could be defined as follows:

$$Z_{mn} = \frac{V_m}{I_n} | I_{k \neq n} \tag{4.1}$$

In practical operation, by injecting a unitary current as an excitation source, I at the nth point, and a unitary current excitation source, $-I$ (equivalent to a sink current I) at an mth point and calculating the resulting voltage at the probe point, we can get the direct impedance Z_{mn} between the m and n contacts.

If we can get the resulting potential distribution in the substrate caused by the unitary current, we could easily get the Z-parameters by (4.1)

Under quasi-static conditions, the potential over the substrate satisfies Laplace's equation:

$$\nabla^2 \varphi(x, y, z) = 0 \tag{4.2}$$

It can also be written as follows:

$$\left(\frac{\partial^2}{\partial x^2} + \frac{\partial^2}{\partial y^2} + \frac{\partial^2}{\partial z^2} \right) \varphi(x, y, z) = 0 \tag{4.3}$$

Hereafter, for Poisson or heat diffusion equations, the analytical method is, first of all, based on the algorithm of variable separation, leading to solve an Eigen problem.

When assuming that $\varphi = X(x, x') \cdot Y(y, y') \cdot Z(z, z')$, it can be rewritten by defining [82]

$$Z(z, z') = Z'(z, z') \cdot \cos\left(\frac{m \cdot \pi \cdot x'}{a} \right) \cdot \cos\left(\frac{n \cdot \pi \cdot y'}{b} \right) \tag{4.4}$$

We get, for a potential due to a point charge, a simple equation:

$$\frac{ab}{4} \cdot \left[\frac{d^2 Z}{dz^2} - \gamma_{mn}^2 \cdot Z \right] = -\frac{\delta(z - z')}{\varepsilon_N} \text{ with } \gamma_{mn} = \sqrt{\left(\frac{m \cdot \pi}{a} \right)^2 + \left(\frac{n \cdot \pi}{b} \right)^2} \tag{4.5}$$

For $z \neq z'$, $\delta(z - z') = 0$. The above equation has a well-known general solution:

$$Z' = A \cdot e^{-\gamma_{mn} \cdot (d+z)} + B \cdot e^{\gamma_{mn} \cdot (d+z)} \tag{4.6}$$

This equation also invokes a transmitted wave and a reflected one.

For instance, for the case that both the point charge and the point of observation are in the same dielectric layer, on the surface with $z = z' = 0$, the Green function then changes to:

$$\varnothing\left(x,y;x',y'\right)_{z=z'=0}\left(G_0\right)_{z=z'=0}$$
$$=\left(\sum_{m=0}^{\infty}\sum_{n=0}^{\infty} f_{mn}C_{mn}\cos\left(\frac{m\pi x}{a}\right)\cos\left(\frac{m\pi x'}{a}\right)\cos\left(\frac{n\pi y}{b}\right)\cos\left(\frac{n\pi y'}{b}\right)\right) \tag{4.7}$$

where $C_{mn} = 0$ for $(m = n = 0)$, $C_{mn} = 2$ for $m = 0$ or $n = 0$ but $m \neq n$ and $C_{mn} = 4$ for all other m and n ($m > 0$ and $n > 0$).

The function f_{mn} is given by

$$f_{mn} = \frac{1}{ab\gamma\varepsilon_N}\frac{\beta_N \tanh\left(\gamma_{mn}d\right)+\Gamma_N}{\beta_N+\Gamma_N \tanh\left(\gamma_{mn}d\right)} \tag{4.8}$$

β_N and Γ_N can be computed recursively from:

$$\begin{bmatrix} \beta_k \\ \Gamma_k \end{bmatrix} = \frac{1}{2}\cdot\begin{bmatrix} 1+\dfrac{\varepsilon_{k-1}}{\varepsilon_k} & \left(1-\dfrac{\varepsilon_{k-1}}{\varepsilon_k}\right)e^{2\theta_k} \\ \left(1-\dfrac{\varepsilon_{k-1}}{\varepsilon_k}\right)e^{-2\theta_k} & 1+\dfrac{\varepsilon_{k-1}}{\varepsilon_k} \end{bmatrix} \times \begin{bmatrix} \beta_{k-1} \\ \Gamma_{k-1} \end{bmatrix} \tag{4.9}$$

where

$$\theta_k = \gamma_{mn}\left(d-d_k\right),\ \beta_0 = 1,\ \Gamma_0 = 1 \tag{4.10}$$

Then, after the derivation of some formal mathematical solutions from Poisson's equation, with boundary conditions, we will depict hereafter a more physical (and graspable) approach of the problem, using a TLM method, with some brief recalls. Recall that, in our whole algorithm, from its beginning, we work on the reciprocal space. More precisely, for instance, it means that we also take the Fourier transform of the injected currents on the discretized contacts.

In practice, then, we can choose any shape for these embedded contacts, whereas, in all the few others works, to the best of our knowledge, all the contacts are rectangular, the integration over them being done in the real space. This last point seems to be crucial to us, since we should work with through silicon vias (TSVs) (cylindrical, tronconic, etc.) in our 3D circuits.

A bidimensional spatial Fourier transform in the x- and y-directions is as follows:

$$\frac{\partial^2\Phi}{\partial z^2} - k_x^2\Phi - k_y^2\Phi = 0 \tag{4.11}$$

where Φ is the spatial-frequency domain potential and k_x, k_y are the spatial-frequency variables (m^{-1}).

Since the current density $J_Z = \sigma^* E_Z$ (σ^* is the complex conductivity) and $E_Z = -\frac{\partial\varphi}{\partial z}$, we can write

$$J_z = -\sigma^* \frac{\partial\varphi}{\partial z} \tag{4.12}$$

In the spatial-frequency domain, it changes to:

$$J_z(k_x, k_y, z) = -(\sigma + j\omega\varepsilon)\frac{\partial\varphi(k_x, k_y, z)}{\partial z} \tag{4.13}$$

$$\begin{cases} \dfrac{\partial\varphi}{\partial z} = -\dfrac{J_z}{(\sigma + j\omega\varepsilon)} \\ \dfrac{\partial J_z}{\partial z} = -(\sigma + j\omega\varepsilon)(k_x^2 + k_y^2)\Phi \end{cases} \tag{4.14}$$

These equations indicate the relationship between current density in the z-direction and the potential distribution in the spatial-frequency domain.

- ***Transmission Line Analogy for Multilayered Media***
  In its simplest form, a transmission line is a pair of conductors linking together two electrical systems (source and load, for instance), with forward and return paths; for cases where the return path is floating, a third conductor (or more) is introduced as the grounding shield. For microwaves, they are waveguides. In our case, the propagation of electron-migration (EM) waves and their interferences, through the silicon substrate, is among the most serious obstacles in the steady trend toward integration of the present-day microelectronics. In fact, the TLM method has been established in some cases concerning substrate surfaces; TLM can be seen as a more physical interpretation of the mathematical developments presented above.

  The principal strength of this method seems to be well dedicated to embedded contacts, irrespective of their number and shape.

  Let us consider a plane wave through a multilayered medium, its plane of incidence being parallel to the <*x*, *y*> plane, providing the same boundary conditions as mentioned above.

  Then, we consider the general case in which the line's impedance is not the same as that of the load. A wave-front A hits the load Z_L: a part of energy is absorbed by Z_L, and the remaining energy is reflected; in this case, voltage and current wave-fronts are not in phase. This reflected wave can meet another incident wave-front B.

The direction of the current flow depends on the polarity of the waveform at the time of observation; if two positive directed waveforms (one forward and one reflected) meet, the current waveforms subtract, but the voltage waveforms add. Likewise, if a positive directed waveform meets a negative directed waveform, the current will add and the voltage will subtract.

The expression for the apparent impedance is given as follows:

$$Z = \frac{V_{\text{total}}}{i_{\text{total}}} = \frac{V_F \pm V_R}{i_F \mp i_R} \tag{4.15}$$

Then, the general solution for voltage and current is as follows:

$$V(z) = V_f e^{\gamma z} + V_r e^{-\gamma z} \tag{4.16}$$

$$I(z) = Y_c\left(V_F e^{-\gamma z} - V_r e^{\gamma z}\right) \tag{4.16 bis}$$

where the propagation coefficient γ is the square root of the product Z by Y, the linear impedance and admittance, respectively,

$$\left(\gamma = \alpha + j\beta;\ \alpha: \text{attenuation},\ \beta = 360°/\lambda_{si}\right)$$

Directly deriving from the chain matrix:

$$\begin{bmatrix} V_1 \\ I_1 \end{bmatrix} = \begin{bmatrix} \cosh \gamma L & Z_c \sinh \gamma L \\ Y_c \sinh \gamma L & \cosh \gamma L \end{bmatrix} \begin{bmatrix} V_2 \\ -I_2 \end{bmatrix} \tag{4.17}$$

And from the output branch $V_2 = -Z_L I_2$, we obtain

$$\begin{bmatrix} V_1 \\ I_1 \end{bmatrix} = -I_2 \begin{bmatrix} Z_L \cosh \gamma L & Z_c \sinh \gamma L \\ Z_L Y_c \sinh \gamma L & \cosh \gamma L \end{bmatrix} \begin{bmatrix} 1 \\ 1 \end{bmatrix} \tag{4.17 bis}$$

where $Z_c = 1/Y_c$ (square root of Z and Y ratio) is the characteristic impedance (electric and magnetic fields' modulus ratio) and Z_L is the charge impedance, and L is the distance from the load. We can extract an input impedance, Z_l ($=V_1/I_1$), function of the load impedance Z_{l+1} (l: layer number), as follows:

$$Z_l = Z_c \frac{Z_{l+1} + Z_c \tanh \gamma L}{Z_c + Z_{l+1} \tanh \gamma L} \tag{4.18}$$

This well-known (TLM) equation Z_l is directly related to (4.8), where $Z_c = \beta_N$ and $Z_L = \Gamma_N$.

In fact, restarting from (4.7) and taking into account the limit or boundary conditions—continuity of potential and discontinuity of the electric field if there is a surface charge at the considered layer interface—it is easy to program this iterative solution versus the substrate depth, or the layers of Figure 4.1. In fact, these calculations are made only one time, at each frequency.

In our simulator, using MATLAB® [63], we use a matrix formalism, extending the impedance (Z) and the current (T) transmission from a layer l to its adjacent one starting from a k layer—see (4.8) or 4.18.

$$\begin{bmatrix} Z_{k,l+1} \\ T_{k,l+1} \end{bmatrix} = \begin{bmatrix} A_k & -B_k \\ C_k & -D_k \end{bmatrix} \begin{bmatrix} Z_{k,l} \\ T_{k,l} \end{bmatrix} \tag{4.19}$$

where T addresses the derivation of the current (between the two complementary parallel regions of the substrate).

Based on the transmission line method, a substrate extractor is programmed, shown in Figure 4.2.

The *ABCD* parameters are variously known as chain, cascade, or transmission line parameters. This cascade connection model is very suitable.

As mentioned earlier, all the calculations are done in the reciprocal space. For instance, the potential is derived from the (Kronecker) product of (sub) matrix Z, by a DCT of the matrix of the injected current, accomplished using two consecutive FFTs. In our algorithm, the contact voltage is the mean value of the voltage of the discrete contact elements, calculated via Millman's theorem, as [85].

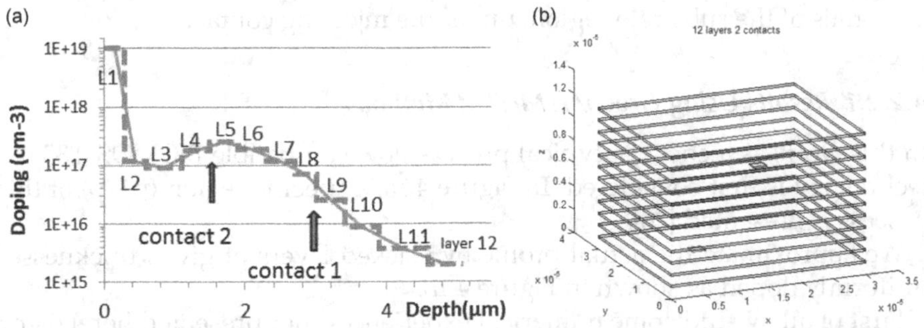

FIGURE 4.1
(a) Specific depth profile of a 35-µm technology (p+/p− region) and (b) 3D schematics of the structure.

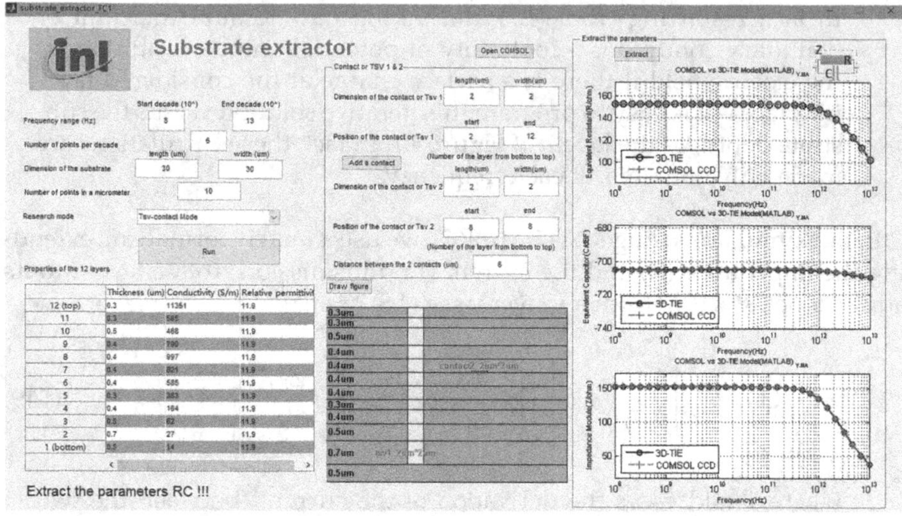

FIGURE 4.2
3D TLE substrate extractor, example: Impedance Z_{17}, capacitance and resistance C_{17} and R_{17} extracted between one contact and one TSV (valid for frequency lower than 20 GHz and of quasi-static assumption).

$$v = \frac{\sum_{n=1}^{N} j\omega c \varphi_n}{\sum_{n=1}^{N} j\omega c} = \frac{1}{N}\sum_{n=1}^{N} \varphi_n \tag{4.20}$$

where φ_n is the substrate voltage at a node and N is the number of contact subareas. Therefore, in our modified algorithm, the contact voltage is then calculated by taking the average <Va> of the potentials of the substrate region, under the injecting contact.

4.2.2.1 Results Using Green/TLM/FEM Methods

In this work, as a check, a typical process flow compatible BiCMOS 0.35 μm technology is first considered. In Figure 4.1a, a specific region (p+/p) of this process is presented [20].

We approximate the actual profile by stacked layers of given thicknesses uniformly doped as shown in Figure 4.1b.

First of all, we did some numerical experiences (not presented here) using COMSOL [43], a well-known Multiphysics (electrical, thermal, and mechanical couplings) simulator, for testing its robustness and accuracy. It is also a dedicated tool for full-wave electromagnetic analysis. It essentially uses

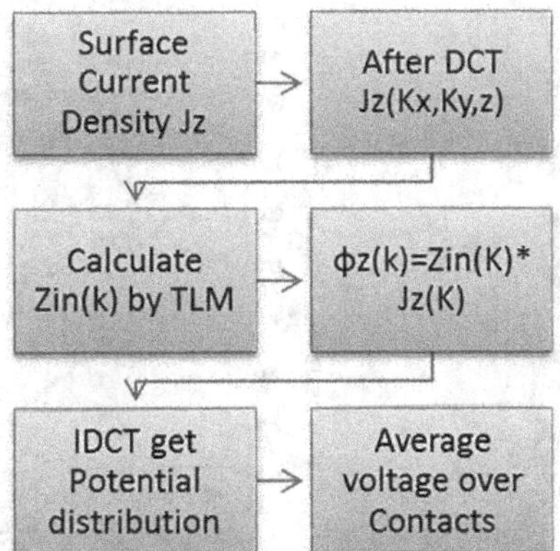

FIGURE 4.3
General steps of 3D-TLE algorithm.

Galerkin-like algorithms; typically, a 3D simulation can use a few ten of minutes or more than 1 h. Figure 4.3 shows the general flowchart of substrate impedance extraction.

- ***Contacts Embedded***
  Considering the general case where the contacts, real or virtual, are totally embedded in Figure 4.1. Then, the substrate can be seen as an actual parallel connection of two separated parts (current derivation). The current is injected on a contact of an *i*th layer, so the substrate is divided in two parts, from this layer.

  Now, it seems to be easy to calculate any transfer impedance by TLM. Possibly embedded in the substrate, contacts can be introduced into any layer; they can be real (e.g., metal like) or virtual (e.g., probe), of any shape.

  The surface die is typically 30 μm × 30 μm, with $M = N = 300$; surface contact: 20 by 20 points (the calculation points, $M \times N$, are equidistributed).

  We present in Figure 4.4a and b the voltage map and the impedance module, respectively, between some C_i and C_j contacts, at the interfaces L_3/L_4 and L_7/L_8. It is typical of a Lorentzian curve ($R//C$, with $R \cdot C = \rho\varepsilon$). The comparison between TLM/Green and COMSOL is quite good. Moreover, in Figure 4.4, surface potentials, from our 3D transmission line extractor (TLE) simulator Figure 4.4a and COMSOL Figure 4.4b for a 12-layer substrate, at a frequency of 100 GHz, are quite similar. Still, a very good accordance was shown between FEM and TLM. Since calculations are made with current source, we get voltage variation under contacts (Figure 4.5).

  From the above results, one can find that the comparison between TLM/Green and COMSOL is quite good. The proposal

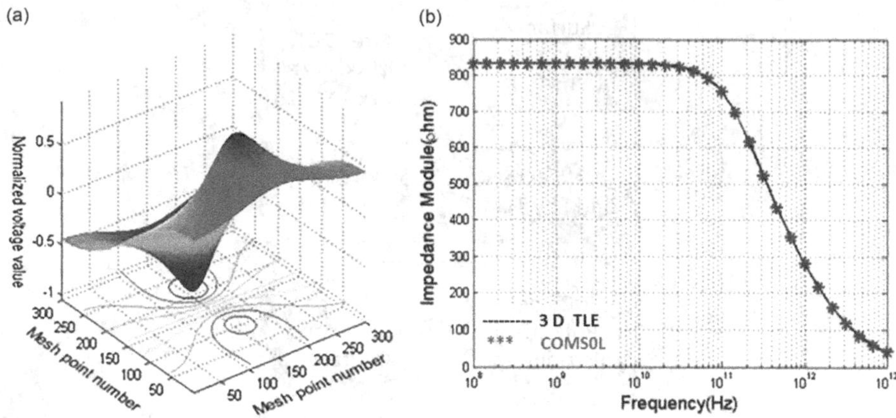

FIGURE 4.4
(a) Voltage map (layer interfaces: L3/L4). (b) Impedance module of Figure 4.1b.

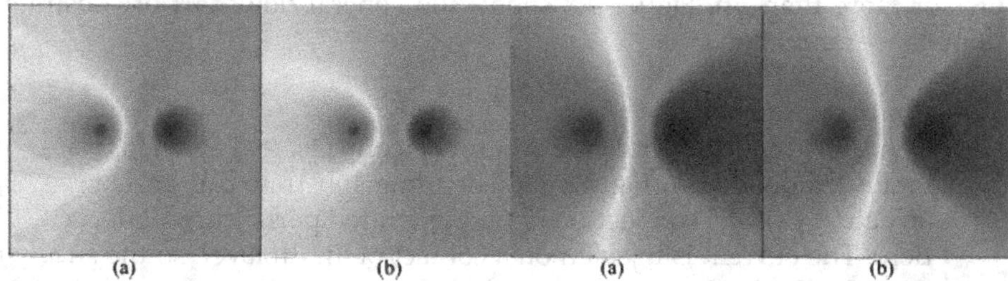

FIGURE 4.5
Frequency = 100 GHz. (1) Left: top surface potential from 3D-TLE (a) and COMSOL (b) for 12 layers substrate. (2) Right (a, b): potential map for contacts embedded in layer four and layer 8 (see Figure 4.1).

method's calculated results agree well with the COMSOL simulation ones, while our method uses comparatively less time (roughly 1/50–1/100). But have in mind that the results to be considered are below ~20 GHz, in the quasi-static regime.

Regarding the potential distribution of the two-contact model in Figure 4.5a, one can find that the potential on the contact surface is not uniform (high-voltage region). The non-uniform potential is caused by an injection of uniform current density instead of a uniform voltage. This non-uniform potential feature makes 3D-TLE very profitable when one calculates the impedance of two resistive regions rather than two metal contacts, for example, the oxide-insulated conductive regions (e.g., pads and coplanar waveguide) and junction insulated well regions (e.g., n-well regions over a p-type substrate). In these cases, the substrate plates could not be considered as an equipotential plate.

The simulation results obtained with Green/TLM implemented in MATLAB and compared with COMSOL are found to be practically equal. Indeed, we choose, in MATLAB, layers as perfect parallelepipeds (uniform thickness layers).

We are quite aware that the quasi-electrostatic modeling framework cannot work so perfectly up to 10 THz (versus experiments), which implies a wavelength of roughly tens of micrometers in Si. This wavelength is of the ten orders small compared to the contact-to-contact distance (3 µm) used; in this work we use quasi-static assumption, not the full-wave Maxwell equations.

Figure 4.6 shows the frequency up to 10 THz, because we want to find out how to simulate parasitic effects at very high frequency. The typical RC model still works for substrate in our case with validation of FEM results, although it is no longer valid in a physical point of view; we should use a full-wave method for the very high frequency (>15 GHz). But it's still a way to compare FEM method (COMSOL) and our calculations. Moreover, for TSV cases, we should be more careful; we want to establish a valid compact model for frequency >15 GHz, so that is why we do the optimization for TSV compact model, because at high frequency (>15 GHz), if we keep the same RLCG model, by adjusting their values, we can still fit with the measurement data, up to 10/15 GHz.

Currently, we did not explicitly take into account the permeability; it is not realistic to go beyond 15 GHz in this analysis (that is not so bad); but we think that it was interesting to compare the FEM in order to test the robustness of our algorithm.

- ***Alteration of the Contacts' Shape and their Number***
  Unlike the pioneered method, based on Green's functions [82] that calculate only for rectangular contacts, our method can be used for

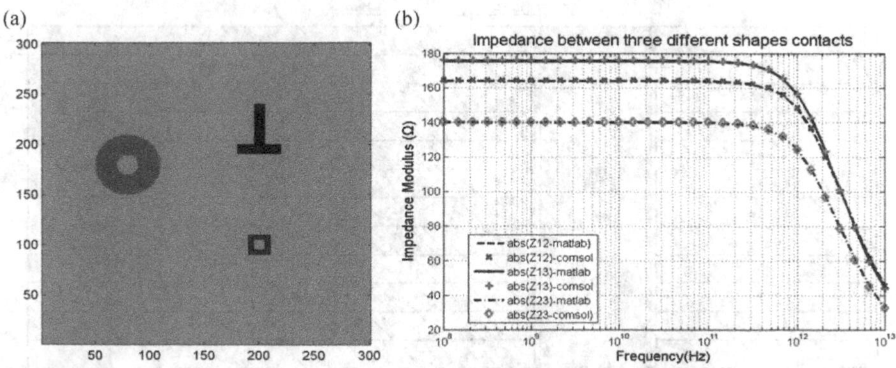

FIGURE 4.6
(a) Some different shapes of contacts, which can be apprehended by 3D TLE (b) Impedances' comparison with COMSOL.

any contact shape: circle, concentric circles, rectangle, loop, T-shape, etc. Figure 4.6 shows some examples from our 3D TLE simulator. We test it by comparing with the results from COMSOL.

First, we apply our model, with three different contacts in Figure 4.6a.

All the three contacts are on the top surface of the substrate. The square contact is contact 1, the T-shape contact is contact 2, and the circle one is contact 3.

The impedance modules between these three contacts, calculated from 3D-TLE and COMSOL, are shown together in Figure 4.6b, in which Z_{12} at the top is the impedance between contacts 1 and 2;

We can see that these results, from the proposed methods, agree very well with the results calculated from COMSOL.

After getting the oxide capacitance, the capacitance effect could be added in 3D-IE program automatically by our following algorithm.

In order to explain how it works, we assume a specific structure as in Figure 4.7.

The matrix G is the admittance matrix which is calculated by 3D-IE. In this specific structure, it has eight rows and eight columns, because we have 1 TSV and seven contacts. If we get the G matrix for a system, we can get the impedance between any two nodes. So the question is how to deduce the new admittance matrix with inserting capacity C_{ox} from the original admittance matrix of the system. The TSV is node 8 which the oxide (capacitance C_{ox}) is added.

This is a structure with 1 TSV and seven contacts. The TSV is surrounding by the oxide coating. The potentials of every contact and TSV are defined in Figure 4.8.

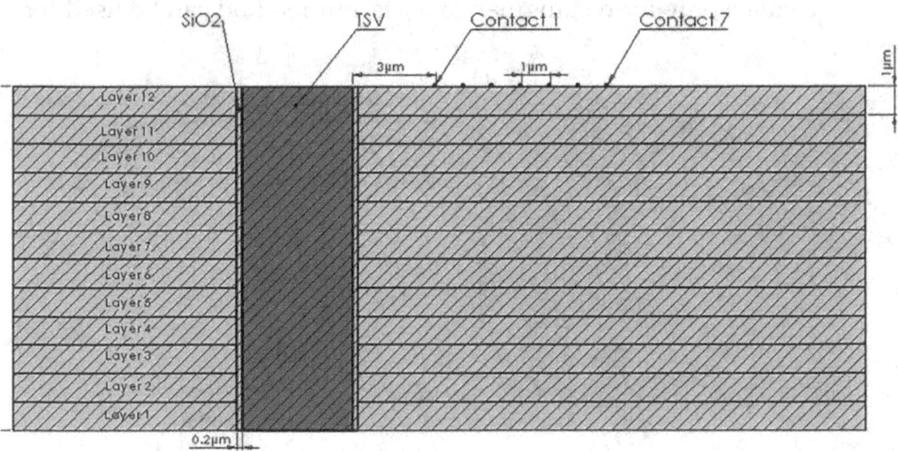

FIGURE 4.7
A specific structure with one TSV (with oxide layer) and seven contacts.

The potentials of contacts are V_1–V_7 and the outside surface potential of TSV coating is V_8 and inside the oxide coating is V_{8c}. Without oxide coating, V_{8c} should equal to V_8.

Before adding the oxide coating, we have:

$$[I]=[G][V] \tag{4.21}$$

i.e.,

$$\begin{bmatrix} I_1 \\ \vdots \\ I_8 \end{bmatrix} = \begin{bmatrix} G_{11} & \cdots & G_{18} \\ \vdots & \ddots & \vdots \\ G_{81} & \cdots & G_{88} \end{bmatrix} \begin{bmatrix} V_1 \\ \vdots \\ V_8 \end{bmatrix} \tag{4.22}$$

After adding oxide coating, the potential changed but the current keep the same value, so we should have:

$$[I]=[G_c][V_c] \tag{4.23}$$

i.e.,

$$\begin{bmatrix} I_1 \\ \vdots \\ I_8 \end{bmatrix} = \begin{bmatrix} G_{11c} & \cdots & G_{18c} \\ \vdots & \ddots & \vdots \\ G_{81c} & \cdots & G_{88c} \end{bmatrix} \begin{bmatrix} V_{1c} \\ \vdots \\ V_{8c} \end{bmatrix} \tag{4.24}$$

From Figure 4.8, $I_8 = \dfrac{V_{8c} - V_8}{Z_s}$ where $Z_s = \dfrac{1}{j\omega c}$

So,

$$V_8 = V_{8c} - I_8 Z_s \tag{4.25}$$

From equation 4.25

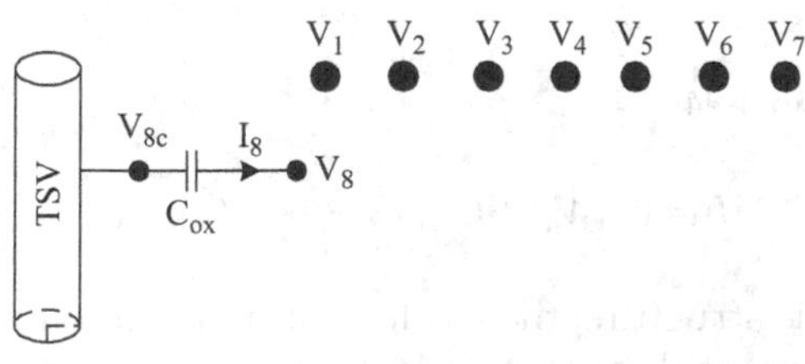

FIGURE 4.8
Abstract model for 1 TSV and seven contacts.

$$
\begin{aligned}
I_8 &= G_{81}V_1 + G_{82}V_2 + \cdots + G_{88}V_8 \\
&= G_{81}V_1 + G_{82}V_2 + \cdots + G_{88}\left(V_{8c} - I_8 Z_s\right) \\
&= G_{81}V_1 + G_{82}V_2 + \cdots + G_{88}V_{8c} - G_{88}I_8 Z_s
\end{aligned} \tag{4.26}
$$

So,

$$
I_8 + G_{88}I_8 Z_s = G_{81}V_1 + G_{82}V_2 + \cdots + G_{88}V_{8c}
$$

$$
I_8\left(1 + G_{88}Z_s\right) = G_{81}V_1 + G_{82}V_2 + \cdots + G_{88}V_{8c}
$$

$$
I_8 = \frac{1}{\left(1 + G_{88}Z_s\right)}\left[G_{81}V_1 + G_{82}V_2 + \cdots + G_{88}V_{8c}\right] \tag{4.27}
$$

Firstly, we deduce the formula for the first row of admittance matrix G. From equation 4.24, we obtain

$$
\begin{aligned}
I_1 &= G_{11}V_1 + G_{12}V_2 + \cdots + G_{18}V_8 \\
&= G_{11}V_1 + G_{12}V_2 + \cdots + G_{18}\left(V_{8c} - I_8 Z_s\right) \\
&= G_{11}V_1 + G_{12}V_2 + \cdots + G_{18}V_{8c} - G_{18}I_8 Z_s
\end{aligned} \tag{4.28}
$$

Substitute equation 4.27 into equation 4.28

$$
\begin{aligned}
I_1 &= G_{11}V_1 + G_{12}V_2 + \cdots + G_{18}V_{8c} - G_{18}Z_s\left(\frac{1}{\left(1 + G_{88}Z_s\right)}\left[G_{81}V_1 + G_{82}V_2 + \cdots + G_{88}V_{8c}\right]\right) \\
&= \left(G_{11} - \frac{1}{\left(1 + G_{88}Z_s\right)}G_{18}G_{81}Z_s\right)V_1 + \left(G_{12} - \frac{1}{\left(1 + G_{88}Z_s\right)}G_{18}G_{82}Z_s\right)V_2 + \cdots \\
&\quad + \left(G_{18} - \frac{1}{\left(1 + G_{88}Z_s\right)}G_{18}G_{88}Z_s\right)V_{8c}
\end{aligned} \tag{4.29}
$$

From equation 4.24:

$$
I_1 = G_{11c}V_{1c} + G_{12c}V_{2c} + \cdots + G_{18c}V_{8c} \tag{4.30}
$$

In our specific structure, the oxide coating is only added to the TSV surface; so we still have $V_1 = V_{1c}, V_2 = V_{2c}, \ldots, V_7 = V_{7c}$. Comparing equations 4.29 and 4.30, one can obtain

$$G_{1jc} = G_{1j} - \frac{1}{(1+G_{88}Z_s)} G_{18}G_{8j}Z_s, \text{ where } (j = 1 \cdots 8) \tag{4.31}$$

Secondly, we deduce a formula for the second row of admittance matrix G. Similarly, from equation 4.22, we can obtain

$$\begin{aligned} I_2 &= G_{21}V_1 + G_{22}V_2 + \cdots + G_{28}V_8 \\ &= G_{21}V_1 + G_{22}V_2 + \cdots + G_{28}(V_{8c} - I_8Z_s) \\ &= G_{21}V_1 + G_{22}V_2 + \cdots + G_{28}V_{8c} - G_{28}I_8Z_s \end{aligned} \tag{4.32}$$

Substituting equation 4.27 into 4.32 yields

$$\begin{aligned} I_2 &= G_{21}V_1 + G_{22}V_2 + \cdots + G_{28}V_{8c} - G_{28}Z_s\left(\frac{1}{(1+G_{88}Z_s)}[G_{81}V_1 + G_{82}V_2 + \cdots + G_{88}V_{8c}]\right) \\ &= \left(G_{21} - \frac{1}{(1+G_{88}Z_s)}G_{28}G_{81}Z_s\right)V_1 \\ &+ \left(G_{22} - \frac{1}{(1+G_{88}Z_s)}G_{28}G_{82}Z_s\right)V_2 + \cdots + \left(G_{28} - \frac{1}{(1+G_{88}Z_s)}G_{28}G_{88}Z_s\right)V_{8c} \end{aligned} \tag{4.33}$$

From equation 4.24, we have

$$I_2 = G_{21c}V_{1c} + G_{22c}V_{2c} + \cdots + G_{28c}V_{8c} \tag{4.34}$$

As before, $V_1 = V_{1c}, V_2 = V_{2c}, \ldots, V_7 = V_{7c}$. Comparing equations 4.33 and 4.34,
we obtain

$$G_{2jc} = G_{2j} - \frac{1}{(1+G_{88}Z_s)} G_{28}G_{8j}Z_s, \text{ where } (j = 1 \cdots 8) \tag{4.35}$$

The steps for deducing the formulas for rows 3 to 7 are similar, now for row 8,

$$I_8 = G_{81c}V_{1c} + G_{82c}V_{2c} + \cdots + G_{88c}V_{8c} \tag{4.36}$$

Due to $V_1 = V_{1c}, V_2 = V_{2c}, \ldots, V_7 = V_{7c}$, so equation 4.27 could rewritten as follows:

$$I_8 = \frac{1}{(1+G_{88}Z_s)}[G_{81}V_{1c} + G_{82}V_{2c} + \cdots + G_{88}V_{8c}] \tag{4.37}$$

Comparing equations 4.36 and 4.37, one can obtain

$$G_{81c} = \frac{1}{(1+G_{88}Z_s)}G_{81}, G_{82c} = \frac{1}{(1+G_{88}Z_s)}G_{82}, \ldots, G_{88c} = \frac{1}{(1+G_{88}Z_s)}G_{88} \tag{4.38}$$

So one has

$$G_{8jc} = \frac{1}{(1+G_{88}Z_s)}G_{8j}, \text{ where } (j = 1\ldots8) \tag{4.39}$$

Equation 4.39 can also be written as follows:

$$G_{8jc} = \frac{1}{(1+G_{88}Z_s)}G_{8j} = G_{8j}\left(1 - \frac{G_{88}Z_s}{(1+G_{88}Z_s)}\right) = G_{8j} - \frac{1}{(1+G_{88}Z_s)}G_{88}G_{8j}Z_s \tag{4.40}$$

Combining equations 4.31, 4.35, and 4.40 yields

$$G_{ijc} = G_{ij} - \frac{1}{(1+G_{88}Z_s)}G_{i8}G_{8j}Z_s \tag{4.41}$$

In this specific structure, the oxide capacitance C_{ox} is added to the node 8 (TSV); if the oxide coating is added to the other nodes (ex. node m) of the structure, we can deduce the total inferential reasoning formula as follows:

$$G_{ijc} = G_{ij} - \frac{1}{(1+G_{mm}Z_s)}G_{im}G_{mj}Z_s \tag{4.42}$$

- ***Model and verification***
  Thus, from the above discussion we can see that adding an oxide coating to a contact (or TSV) is only a transform of admittance matrix form G_{ij} to G_{ijc} using equation 4.42. After adding the oxide coating to the TSV, the new equivalent circuit between TSV and contacts is shown in Figure 4.9.

  In order to verify the result of our 3D-IE, a model is constructed in COMSOL as shown in Figure 4.10; the blue coating outside the TSV is oxide layer. The distance between TSV and contact is 6 μm. The TSV and contact are both square 2 μm × 2 μm; thickness of oxide layer is 0.5 μm outside TSV. The substrate structure and property are the same as before.

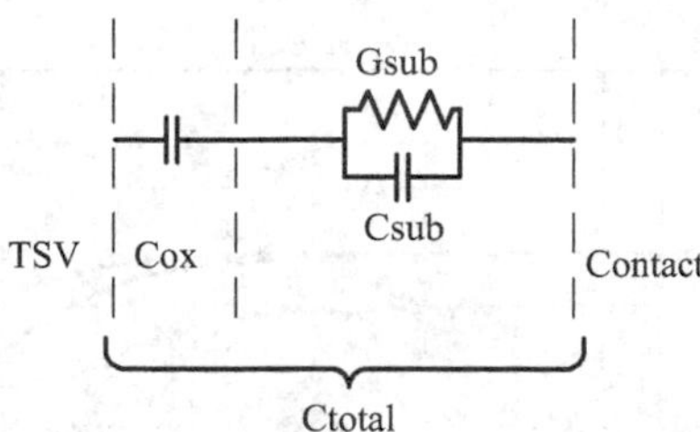

FIGURE 4.9
New model for the capacitance total between TSV and contact.

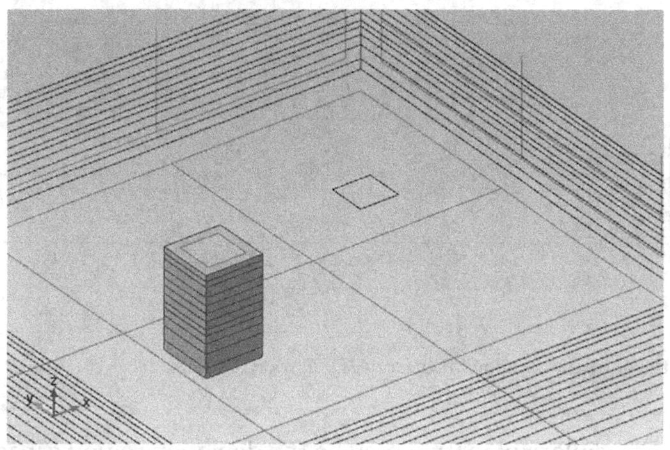

FIGURE 4.10
Model for verification of adding oxide coating.

As shown in Figure 4.10, the total capacitance between TSV and contact is extracted by 3D-IE and COMSOL separately. The C_{total} is a combination of the capacitance oxide C_{ox} and capacitance substrate $C_{sub.}$

In 3D-IE, the following two different methods are used:

1. Treat the TSV as a combination of some pieces and add C_{ox} to each part of the TSV.
2. Combine all the TSV parts to be a whole TSV and add a C_{ox} totally directly.

The results of C_{total} calculated by 3D-IE and COMSOL are shown in Figure 4.11.

From Figure 4.11 one can find, as compared with the result of COMSO, that method A is more accurate in high frequency than method B; but they present both a lack of accuracy.

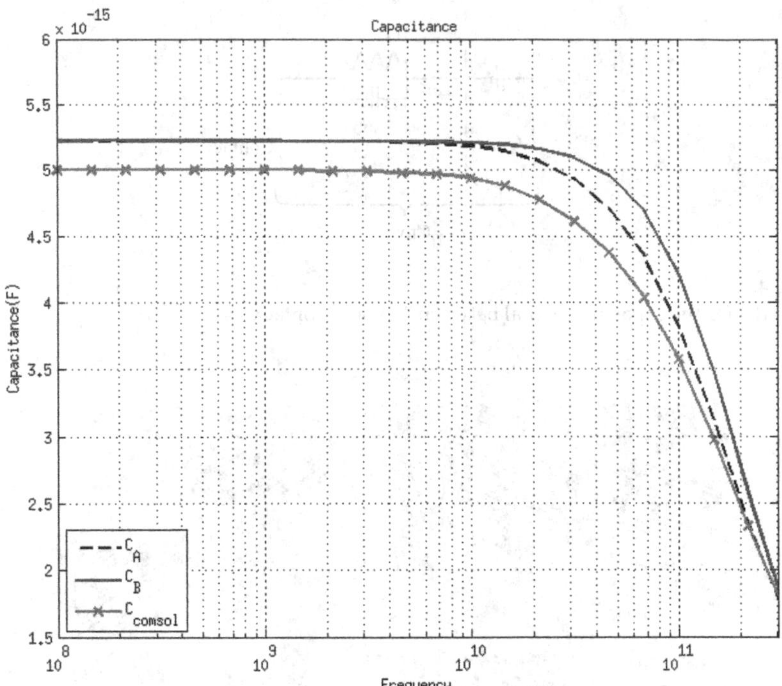

FIGURE 4.11
Comparison the equivalent capacitance from 3D IE and COMSOL.

One possible explanation may be the depletion region capacitance which is in series with the capacitance oxide.

The depletion region is also called depletion layer, depletion zone, junction region or the space charge region; it is an insulating region within a conductive, doped semiconductor material where the mobile charge carriers have diffused away, or have been forced away by an electric field. The only elements left in the depletion region are ionized donor or acceptor impurities.

A full derivation steps for the depletion width could be found in [76]. The derivation is based on solving the Poisson equation (Figure 4.12).

- ***Model via-contact***
  In order to test our extractor's accuracy, we compare our simulation results with ones coming from COMSOL.

  Here (Figure 4.13), a TSV of 5.1 μm side occupies the 12 layers of the substrate.

  The pitch via-contact is set to 6 μm and the frequency varies from $1e^{8}$ to $1e^{13}$ Hz. We set the surface of the via as a port and input 1A current and –1A at the other contact's entirety. The frequency

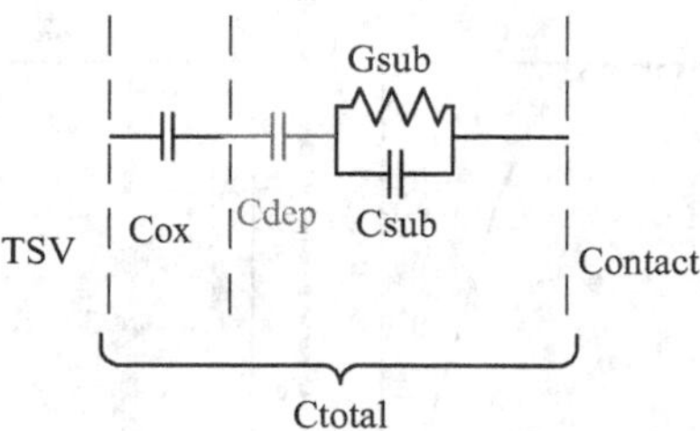

FIGURE 4.12
A possible model for the substrate with depletion region.

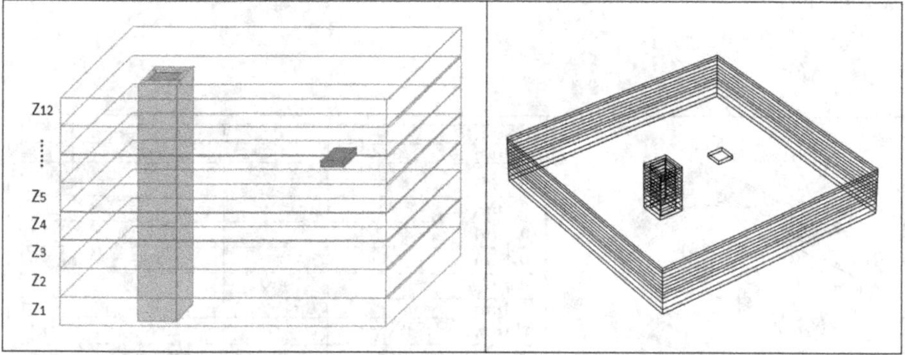

FIGURE 4.13
Via-contact contact.

RC-parameters results, from the two simulators, are shown in Figure 4.14 They show some good fits. The variation between the two simulators increases versus frequency (essentially at frequencies above 20 GHz, limit of the quasi-static regime; i.e., limit of our model); but even at 10 THz (far from this quasi-static regime), the variation percentage is only 1.4% maximum. Figure 4.14 could be a way to define the quasi-static regime limit frequency f_{QS}. We think that the definition of layers, and their increase, for the TLM model, reducing the thicknesses of each subdomain (layer), pushes back to the high frequencies the validation frequency f_{QS}. Work with a full-wave simulator should conclude this question.

To go one step further, we extract scattering parameters of two via embedded in the substrate—Figure 4.15—where each "piece" of TSV is surrounded by a thin oxide/inducting same capacitance, with an RDL connecting them. Besides, this inter via line (RDL) between the two TSV can be modeled as an equivalent RL circuit, or RLC considering the oxide which surrounds it. Its resistance (R_{line}) and

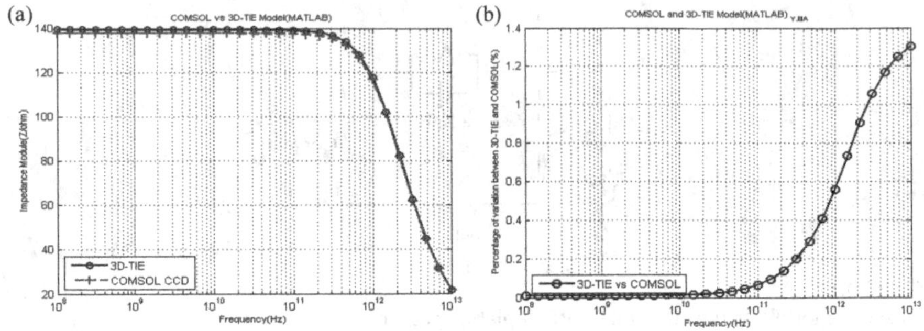

FIGURE 4.14
Via-contact: (a) Comparison between Green and FEM simulation. (b) Error rate between the two methods.

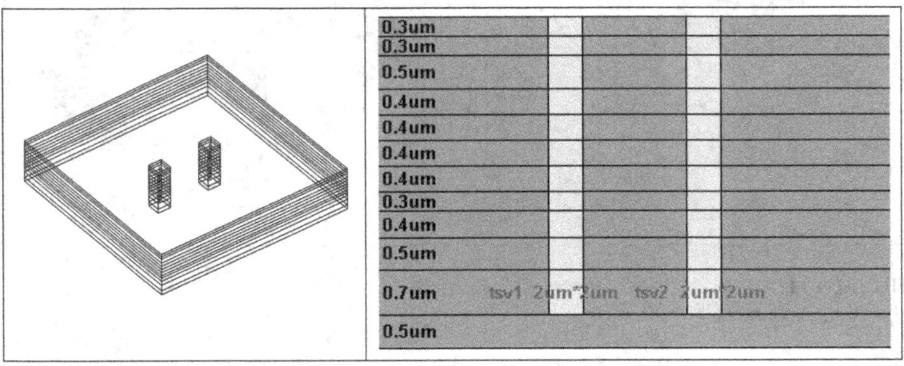

FIGURE 4.15
Via-via model.

inductance (L_{line}) are calculated analytically and, regarding its weak dimensions, can be considered as constant with frequency, up to 20 GHz. We find, with a first draft, some good fits between experiments, numerical simulations, and our analytical ones.

Here our key goal is to explain our methodology; but we should consider, in future trends the capacitance not only induced by the coating oxide but also by a depletion layer in series with this latter one, mainly for a low-doped substrate; for instance, resolving Poisson's equation in cylinder coordinates (in 2D, if the length of the via is very large compared to its diameter).

The pitch is set to 6 μm and the frequency varies from $1e^8$ to $1e^{13}$ Hz. We set the surface of the via as a port and inject uniformly a 1 A current in it and –1 A into the other contact (Figure 4.16).

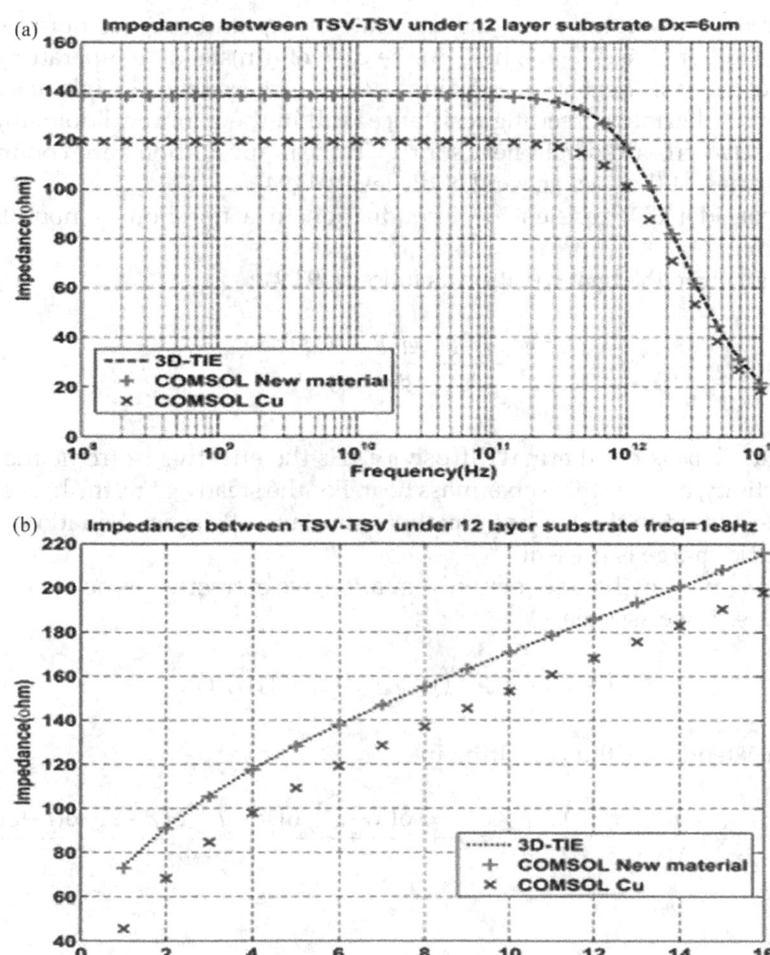

FIGURE 4.16
Via-via: comparison between Green and FEM simulation.

4.3 Heat Equation

A problem commonly encountered in the thermal analysis of electronic packages, or in 3D ICs, is that of thermal spreading resistance. Thermal spreading resistance occurs as heat flows by conduction between a source and a sink, with different cross-sectional areas. One example is that of a multilayer chip. Another one is that of a module package with a thermal grease path between a chip and the module cap. In either case, heat flows from the chip across a given area into the substrate or cap and spreads out to leave across another area. We are, here, in the case of (un)steady temperature due to one or multiple heat sources, using in this last case the principle of superposition. A thermal spreading resistance (still in the reciprocal domain) can be extracted from discrete heat source. Typical applications are cooling of device areas, at the package and at die level [93,94].

We consider 3D transient heat conduction in a non-homogenous finite medium.

We can write the heat equation as follows [93,94]:

$$\frac{d^2T}{dx^2}+\frac{d^2T}{dy^2}+\frac{d^2T}{dz^2}-\frac{1}{\alpha}\frac{dT}{dt}=0 \tag{4.43}$$

where $\alpha = k/\rho c$ is the thermal diffusivity, k is the effective isotropic thermal conductivity, and c is the isobar mass heat. For the steady state, the heat equation is resumed to the Laplace equation, similar to Poisson's equation when no electric charge is present.

In the same way that for Poisson's equation, we extract an associated Green kernel, i.e., if we assume

$$T = X(x,x')\cdot Y(y,y')\cdot Z(z,z')\tau(t,t'), \tag{4.44}$$

we can extend 4.43, thus resulting in

$$\frac{1}{X}\cdot\frac{d^2X}{dx^2}+\frac{1}{Y}\cdot\frac{d^2Y}{dy^2}+\frac{1}{Z}\cdot\frac{d^2Z}{dz^2}=-A^2\frac{\delta(x-x')\cdot\delta(y-y')\cdot\delta(z-z')\cdot\delta(t-t')}{\alpha} \tag{4.45}$$

Then:

$$\frac{d^2X}{dx^2}+\lambda^2X=\frac{d^2Y}{dy^2}+\mu^2Y=\frac{d^2Z}{dz^2}-\gamma^2Z=\frac{d\tau}{dt}-\alpha\left(\lambda^2+\mu^2-\gamma^2\right)\tau=0 \tag{4.46}$$

where λ, μ, and γ are the eigenvalues of (4.23) associated with the Green kernel of this heat equation with:

$$\lambda^2+\mu^2-\gamma^2+\frac{A^2}{\alpha}=0 \tag{4.47}$$

The last equation in 4.45 has an evanescent solution versus time; for an ith layer, we can write

$$\tau_i=\tau_{i0}\cdot\exp(-\left(\lambda_{ix}^2+\mu_{iy}^2-\gamma_{iz}^2\right)\cdot\alpha\cdot t \tag{4.48}$$

Staying static in the reciprocal domain (so, we can also take advantage of the data compression, inherent to the DCT), equation 4.43 becomes:

$$\frac{d^2X}{dx^2}+\frac{d^2Y}{dy^2}+\frac{d^2Z}{dz^2}-\frac{1}{\tau}.\frac{d\tau}{dt}=-\lambda^2X-\mu^2Y+\gamma^2Z-\left(\frac{j}{\alpha}\right)\omega\tau=0\ \left(j^2=-1\right) \tag{4.49}$$

We can define a Fourier coefficients' spreading (transcendental) impedance function as the following ratio:

$$R_{\mathrm{FCS}i}=A_i/B_i \tag{4.50}$$

where A_i and B_i are the Fourier coefficients of the general ith solution for the temperature, in the z-direction perpendicular to the stacked layers

$$\mathrm{FCS}_i=A_i\cdot\cosh(\gamma z)+B_i\cdot\sinh(\gamma z) \tag{4.51}$$

- ***Boundary conditions***
  The continuity of the thermal conductivity at interface layers leads to:

$$k_i\frac{dT_i}{dz_i}=k_{i+1}\frac{dT_{i+1}}{dz_{i+1}}\ (\text{Neumann condition}) \tag{4.52}$$

We have also at interface

$$X_i(t)=X_{i+1}(t),Y_i(t)=Y_{i+1}(t)\ (\text{Dirichlet condition}) \tag{5.53}$$

$$k_i\frac{T_{i,z}}{T_{i+1,z}}=k_{i+1}\frac{R_{\mathrm{FCDS}_{i,z}}}{R_{\mathrm{FCDS}_{i+1,z}}} \tag{5.54}$$

See the analogy with the induction vector; at an interface without charge, we have

$$\varepsilon_i\mathrm{E}_i=\varepsilon_{i+1}\mathrm{E}_{i+1}\ (\mathrm{E}:\text{Electrical field}) \tag{4.55}$$

If (4.51) and (4.53) are verified, it follows:

$$R_{\mathrm{FCS}_{i,z}}\cdot Y_{i(y)}\cdot e^{-(-\gamma_{i,z}^2+\mu_{i,y}^2+\lambda_{i,x}^2)t}=-B_{i,z}\cdot R_{\mathrm{FCS}_{i+1,z}}\cdot Y_{i+1(y)}\cdot e^{-(-\gamma_{i+1,z}^2+\mu_{i+1,y}^2+\lambda_{i+1,x}^2)t} \tag{4.56}$$

$$k_{i,z} \cdot R_{\text{FCS}_{i,z}} \cdot Y_{i(y)} \cdot e^{-(-\gamma_{i,z}^2 + \mu_{i,y}^2 + \lambda_{i,x}^2)t}$$
$$= -k_{i+1,z} \cdot B_{i+1,z} \cdot \gamma\lambda R_{\text{FCS}_{i+1,z}} \cdot Y_{i+1(y)} \cdot e^{-(-\gamma_{i+1,z}^2 + \mu_{i+1,y}^2 + \lambda_{i+1,x}^2)t} \quad (4.56 \text{ bis})$$

We have the same set, 4.56, 4.56 bis, replacing Y with X. Moreover, 4.56 and 4.56 bis are verified if:

$$\frac{B_{i,z}}{B_{i+1,\,z}} = \frac{R_{\text{FCS}_{i+1,z}}}{R_{\text{FCS}_{i,z}}} \quad (4.57)$$

$$\frac{k_{i,z}}{k_{\,i+1,z}} = \frac{R_{\text{FCS}_{i,z}}}{R_{\text{FCS}_{i+1,z}}} \quad (4.58)$$

Note that for different thermal diffusivities or conductivities, in each layer, but with $k_{x,i} = k_{y,i}$ (see isotropy).

$$\gamma_{i+1,z} = \left[\frac{\alpha_i}{\alpha_{i+1}} \cdot \gamma_{i,z}^2 + \left(\frac{\alpha_i}{\alpha_{i+1}} - 1\right) . \left(\gamma_{i,x}^2 + \mu_{i,y}^2\right)\right]^{\frac{1}{2}} \quad (4.59)$$

We obtain:

$$A_{iz} = B_{iz} \cdot R_{\text{FCDS}_{i,z}} \cdot \left(\gamma_{i,z}\right) \quad (4.60)$$

$$R_{\text{FCS}_{i,z}}\left(\gamma_n\right) = \frac{k_{i+1} R_{\text{FCS}_{i+1,z}}\left(\gamma_n\right) + k_i \cdot \tanh\left(\gamma_{i,z} \cdot t_i\right)}{k_i \cdot R_{\text{FCS}_{i+1,z}}\left(\gamma_n\right) \cdot \tanh\left(\gamma_{i,z} \cdot t_i\right)} \quad (4.61)$$

Here, $\gamma_{i,z}$ is the eigenvalue in (4.47); t_i is the thickness of the ith layer.

The above development seems to be important because it does not have the classic form of previous works, especially in this thermal area. Furthermore, we get a shape analogous to the thermal impedance and the recursive electrical impedance specific to the TLM method (produced also by Green kernels). Moreover, to the best of our knowledge, this triple analogy (cf. Chapter 5: noise), working in the reciprocal field, seems innovative.

The impedance associated with the Fourier coefficient ratio of temperature solution is presented in Figure 4.17. Using this low-pass filter, tuning a maximum frequency should permit us, for instance, to scrutinize the thermal flow behavior across each interface layer, for a frequency range domain. In addition, taking inverse Fourier transforms, we could compare the direct calculation of time-dependent heat profile (see FEM).

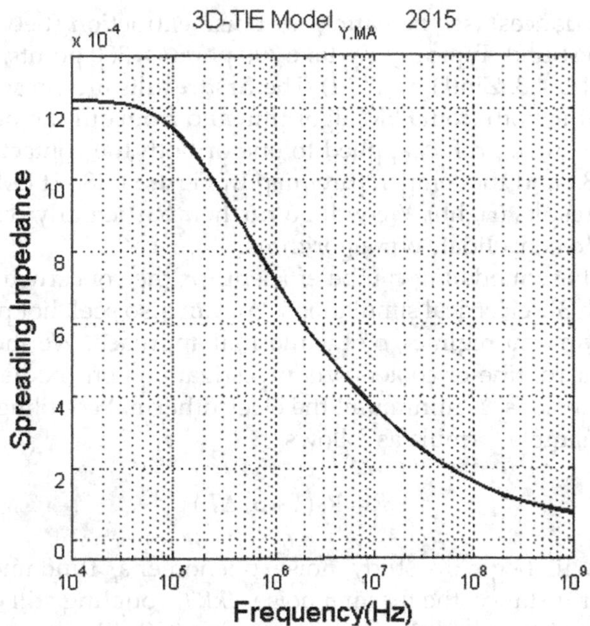

FIGURE 4.17
Temperature Fourier coefficients' spreading impedance.

Finally, in general, the real solution of 4.43 takes the following form:

$$\Delta T\left(x,y,x',y',z,z'\right)=T-T_{\infty}=T_0\cdot\exp\left(-\left(\lambda_{i,x}^2+\mu_{i,y}^2-\gamma_{i,z}^2\right)\cdot\alpha\cdot t\right)\cdot$$

$$\sum_{m=1}^{\infty}\sum_{n=1}^{\infty}\left[\frac{2\left(A_N e^{-\gamma_{mn}(d+Z_l)}+B_N e^{\gamma_{mn}(d+Z_l)}\right)\left(e^{-\gamma_{mn}Z_h}+e^{\gamma_{mn}Z_h}\right)}{ab_{N\gamma_{mn}}\left(B_N e^{d}-A_N e^{-d}\right)}\right]\cos\left(\frac{mx}{a}\right)\cos\left(\frac{ny}{b}\right) \tag{4.62}$$

where T_{∞} is the stationary temperature.

4.3.1 Typical Simulation Results

We built our simulator in order to make an efficient use of the analogies between, first of all, electrical and thermal models. In particular, if all the relevant variables' calculation (voltages/temperature, current density/thermal flow, noise power/current fluctuations) made in the reciprocal domains of space and frequency is done in the main program, we do develop a single special subroutine to calculate the impedance (electrical or thermal, or transfer impedances) (see hereafter) for noise extraction (between any two points into the bulk). Typically, we take 300 points × 300 points, and 12 layers (see Figures 4.1–4.3: Zin(k) by TLM). The first results are presented for temperature coming both from FEM algorithm and from our method, where the temperature source profile, applied to one or both the contacts, is depicted in Figure 4.18a and the temperature map in Figure 4.18b (FEM results (left) and Green's results (right)). We will go further, particularly with numerical experiences, to consolidate our algorithms.

We think these modeling can be efficient for the concerned industry, for instance, with 3D electrical simulations. We can also seek hot points, involving drastic transient regimes, by the thermal approach. We should work on the placement routine of coolers, via investigations on the heat spreading, in three dimensions. A minimum, the electrothermal coupling can be done by the linear simple formula as follows:

$$R = R_0\left(1 + \alpha_0 \Delta T\right) \tag{4.63}$$

Finally, we will begin to study noise phenomena, fundamental for the nanoscale, for instance, the thermal noise, *4kRT*, coupling still electrical and thermal fields, or the generation-recombination (GR)-like one, as well as their coupling (Chapter 5).

4.3.2 Thermal Connection Modeling

After a thermal substrate analysis, we inspect the thermal behavior of TSVs, the key elements of connections embedded in the substrate. 3D IC integration can effectively increase the system's integrity, but as a side effect, the following increase of system power density can cause serious heat dissipation problem [93–95]. It has become a bottleneck for the 3D IC. In this case, how to

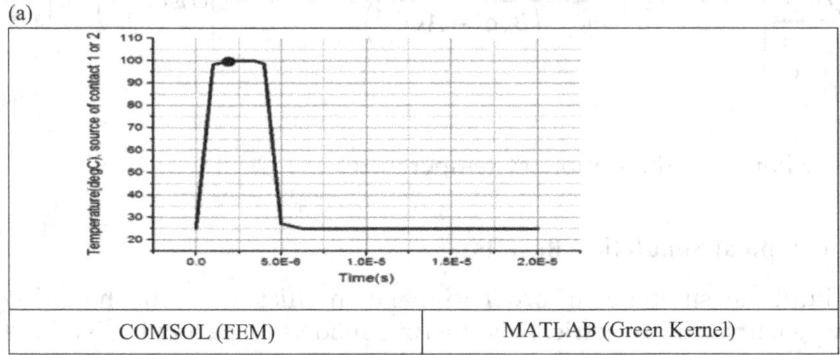

FIGURE 4.18
(a) Temperature profile and (b) temperature maps.

(Continued)

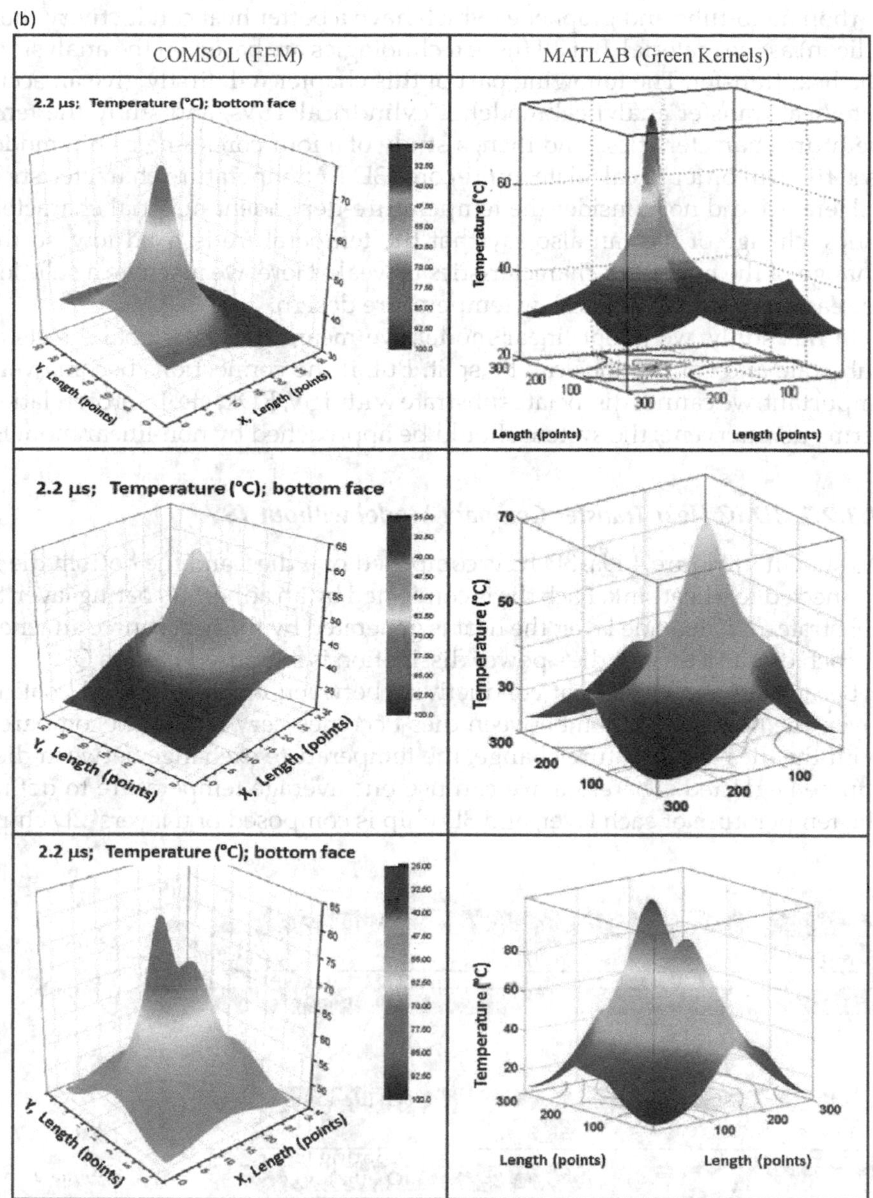

FIGURE 4.18 (CONTINUED)
(a) Temperature profile and (b) temperature maps.

remove heat from the IC circuits and realize an effective heat management becomes the key of the 3D IC technology. In order to solve the heat dissipation problem, researchers from all over the world have proposed some heat management solutions, such as conducting TSV, micro-pumps, using the new carbon nano-tube and graphene (which have a better heat conductivity than silicon) as the material. But all these technologies are based on the analysis of the heat transfer. The following part of this chapter will firstly give an accurate heat transfer analytical model of cylindrical TSVs, and study the temperature characteristics. And then, a study of a four cores single chip model is settled, in order to calculate multi-cores 3D IC temperature characteristics.

Here, we did not consider the temperature-dependent material characteristics' change or we can also say that the temperature is fixed low, so the change of the material's characteristics is weak. Here, we just give a solution for each uniform layer and low temperature design.

In our study, we adopt linear model; we mean that the global effects of substrate and connections can be split. But, if the connections become very important, we cannot dissociate substrate with TSV, RDL, etc. Intercorrelation terms do intervene; the system should be approached by non-linear models.

4.3.2.1 3D IC Heat Transfer Compact Model without TSV

As shown in Figure 4.19a, 3D IC is composed of n dies, and the bottom die is connected to a heat sink. Each die is connected by an adhesive coating layer, Si substrate and an oxide layer; the heat is generated by the heat source attached at each die, and the *n*th die's power dissipation is supposed here as Q_n.

Because of the weak heat conductivity between oxide layer and coating layer, the heat conduction between dies becomes very hard, and compared with the dies' temperature change, the temperature exchange between dies can be neglected. Therefore, we can use one average temperature to define the temperature of each layer, and 3D chip is composed of n layers' 2D chips

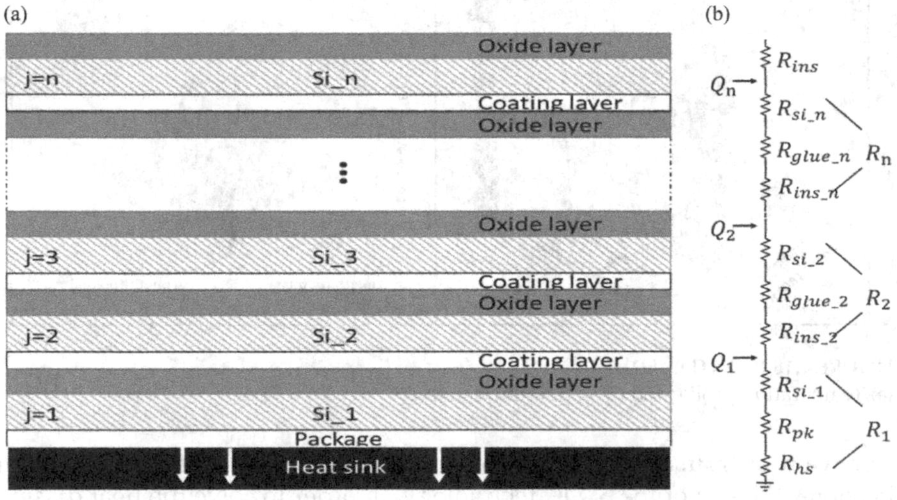

FIGURE 4.19
(a) Schematic of 3D IC structure (without Vias) and (b) heat transfer model.

with different temperatures. According to the Fourier heat flow analysis, using the heat flow as an analogy of the current, temperature as an analogy of the voltage, heat resistance as an analogy of the resistance, we can get the jth layer's temperature [95].

$$T_j = \sum_{i=1}^{j}\left[R_i\left(\sum_{m=i}^{n} Q_m\right)\right] \tag{4.64}$$

Here, n is the total layer number, R_m and Q_m are, respectively, the mth layer's heat resistance and power dissipation [96,97].

Assuming that the each layer's power dissipation is the same, all Q, the first layer's thermal resistance is R_1, and the rest layers' thermal resistances are the same R, so for the n layers 3D IC, the top layer' temperature can be expressed as follows:

$$T_n = Q\left[\frac{R}{2}n^2 + \left(R_1 - \frac{R}{2}\right)n\right] \tag{4.65}$$

From Figure 4.19b, we can see that the first die's thermal resistance R_1 includes the thermal resistance of S_{i_1} substrate, package, and heat sink; the other dies' thermal resistance R includes the thermal resistance of S_i substrate, coating layer, and oxide layer. The thermal resistance is defined as $R = l/(kS)$, l is the heat flow length, k is the heat conductivity, S is the cross-section surface of heat flow; Q represents the power dissipation, so R_1 and R represent, respectively [3,4]:

$$R_1 = R_{si} + R_{pk} + R_{hs} = \frac{l_{si}}{k_{si}S} + R_{pk} + R_{hs} \tag{4.66}$$

$$R = R_{si} + R_{glue} + R_{ins} = \frac{l_{si}}{k_{si}S} + \frac{l_{glue}}{k_{glue}S} + \frac{l_{ins}}{k_{ins}S} \tag{4.67}$$

where R_{si}, R_{pk}, R_{hs}, R_{glue}, and R_{ins} represent, respectively, thermal resistance of each dies' silicon substrate, package, heat sink, coating layer, and insulation layer, l_{si}, l_{glue}, l_{ins} represent, respectively, the thickness of silicon substrate, coating layer, and insulation layer, k_{si}, k_{glue}, k_{ins} represent the thermal conductivity of silicon substrate, coating layer, and insulation layer, S is the chip's surface.

4.3.2.2 3D IC Heat Transfer Compact Model Considering TSV

In the 3D IC, TSV generally is filled with copper, because of its high thermal conductivity and strong heat dissipation power. If we neglect TSVs in the heat transfer model, it will lead to an over-estimation of the chip temperature; so it's necessary to consider TSV parasitic when analyzing 3D IC heat characteristics. Here we adopt a 3D IC top layer analytical model by considering TSVs.

After adding TSVs in the 3D IC, the structure will change to Figure 4.20a,b: heat transfer model.

Assuming that the total surface of each layer in the IC is S_{TSV}, the total chip surface is S, and TSV surface ratio is r, defined as follows:

$$r = \frac{S_{\text{TSV}}}{S} \tag{4.68}$$

When considering TSVs, the layer's thermal resistance will change to the parallel of TSV thermal resistance and the original thermal resistance, as shown in Figure 4.20b. Assuming chip's surface stays unchanged, by adding TSV, the other parts' surface of the layer will be reduced, and that will be $(1 - r)S$. So, the first layer's thermal resistance will change from R_1 to R_1'

$$R_1' = \frac{l_{\text{si}}}{k_{\text{si}}(1-r)S} \Bigg\| R_{\text{TSV}} + R_{\text{pk}} + R_{\text{hs}} = \frac{l_{\text{si}}}{k_{\text{si}}(1-r)S} \Bigg\| \frac{l_{\text{si}}}{k_{\text{TSV}} rS} + R_{\text{pk}} + R_{\text{hs}} \tag{4.69}$$

The other layers' thermal resistance will change from R to R':

$$R_1' = \frac{l_{\text{si}}}{k_{\text{si}}(1-r)S} \Bigg\| R_{\text{TSV}} + R_{\text{pk}} + R_{\text{hs}} = \frac{l_{\text{si}}}{k_{\text{si}}(1-r)S} \Bigg\| \frac{l_{\text{si}}}{k_{\text{TSV}} rS} + R_{\text{pk}} + R_{\text{hs}} \tag{4.70}$$

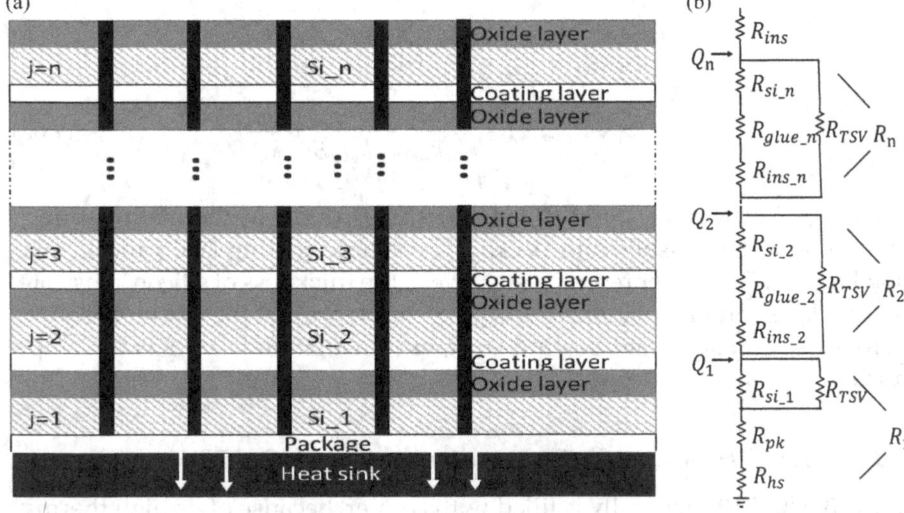

FIGURE 4.20
(a) Schematic of 3D IC structure with TSVs and (b) heat transfer model.

So, after considering TSV in the 3D IC, top layer temperature analytical model can be written as follows [3,4]:

$$T_n = Q\left[\frac{\left(\left(\frac{l_{si}}{k_{si}}+\frac{l_{glue}}{k_{glue}}+\frac{l_{ins}}{k_{ins}}\right)\frac{1}{(1-r)S}\,\|\,\frac{l_{si}+l_{glue}+l_{ins}}{k_{TSV}rS}\right)}{2}n^2 + \left(\left(\frac{l_{si}}{k_{si}(1-r)S}\,\|\,\frac{l_{si}}{k_{TSV}rS}+R_{pk}+R_{hs}\right) - \left(\frac{\left(\frac{l_{si}}{k_{si}}+\frac{l_{glue}}{k_{glue}}+\frac{l_{ins}}{k_{ins}}\right)\frac{1}{(1-r)S}\,\|\,\frac{l_{si}+l_{glue}+l_{ins}}{k_{TSV}rS}}{2}\right)\right)n\right] \quad (4.71)$$

Here, we use an 8 dies 3D IC model, and we choose MATLAB to do the simulation. Table 4.1 gives the model parameters.

Through the MATLAB simulation, we can get the TSV top layer's temperature's change with the IC layer numbers, as shown in Figure 4.21. We can see that, with the increase of number of layers, the chip top surface temperature increases linearly, the three colored lines represent r equals to 0.0001, 0.001, 0.01 and the black dotted line represents the dies without TSV. For the case without TSV, the top surface temperature can reach 427K, but after considering TSVs, the decrease of temperature is more obvious. The greater the r is, the lower the temperature is. So for an eight-layer 3D IC, when r equals to 0.0001, the temperature of top layer decreases from 427 to 421K, when $r = 0.001$, it decreases to 392K, and when $r = 0.01$, it's 360K; this is because of the heat dissipation effect of TSV.

As shown in Figure 4.20, the 3D IC top layer's temperature changes with TSV surface ratio r and dies' number n. For the same r, with the increase of dies' number n, r has more and more obvious influence to temperature. For a very high n, when we decrease r, at the beginning, the chip's temperature does not change; when r arrives enough small, the temperature T increases heavily.

TABLE 4.1

8 Dies 3D IC Model's Parameters

Q (W)	L_{si} (μm)	L_{glue} (μm)	L_{ins} (μm)	k_{si} (W/mK)	k_{glue} (W/mK)	k_{TSV} (W/mK)	k_{ins} (W/mK)	R_{hs} (K/W)	R_{pk} (K/W)	S (mm²)
2	50	2	8.8	150	0.25	390	0.07	2	20	10 * 10

Figure 4.21 shows the temperature changes when $n = 8$. When r equals to 0.0001, 3D IC top surface's temperature is 421 K; with the increase of r, the temperature decreases. When r equals to 0.005, the temperature decreases to 366K, the decrease can reach 55 K, about 13%; when r equals to 0.01, to compare with r equals to 0.005, the decrease is 6 K, about 1.6%. When r equals to 0.015, the temperature is 357 K, compared with $r = 0.01$, the decrease has only 3 K, about 0.8%, and with the augmentation of r, the temperature doesn't have a significant change. Thus, in the 3D IC design, for $n = 8$, the best surface ratio of TSVs is 0.5%~1%. It can be adjusted with different demands. If the chip does not need so many interconnection signal line, the other TSVs can be changed to thermal via and get a better heat transfer. 3D IC designer can read directly from Figure 4.22, the r value according to their temperature demand. When n equals to other value, r can also be calculated by this way.

But if we want to get every point's temperature of the 3D IC, another 2D heat transfer solver should be used to generate accurately the temperature 2D distribution. Here, we adopt a finite volume method (FVM). The heat diffusion equation is in general used to describe the heat conduction in a chip and calculate the temperature profile (Figure 4.23):

$$\rho c_p \frac{\partial T(\vec{r},t)}{\partial t} = \nabla \cdot \left[k(\vec{r},T) \nabla T(\vec{r},t) \right] + g(\vec{r},t) \tag{4.72}$$

which is subject to the general thermal conduction boundary condition.

$$k(\vec{r},T) \frac{\partial T(\vec{r},t)}{\partial n_i} = h_i \left(T_a - T(\vec{r},t) \right) \tag{4.73}$$

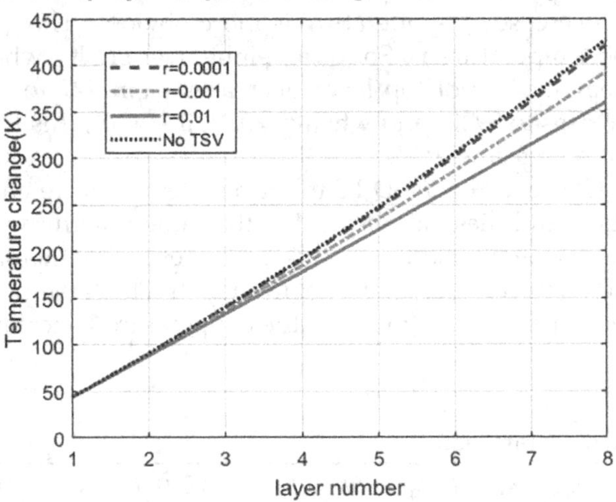

FIGURE 4.21
Different layer number's influence to the 3D IC top layer temperature.

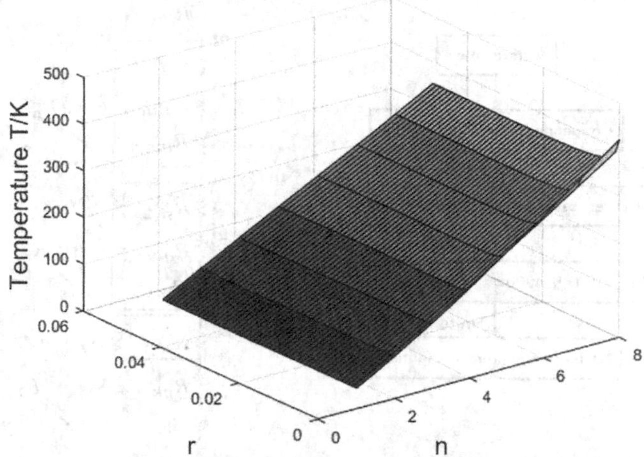

FIGURE 4.22
Surface ratio r and dies' number's influence to 3D IC top layer temperature.

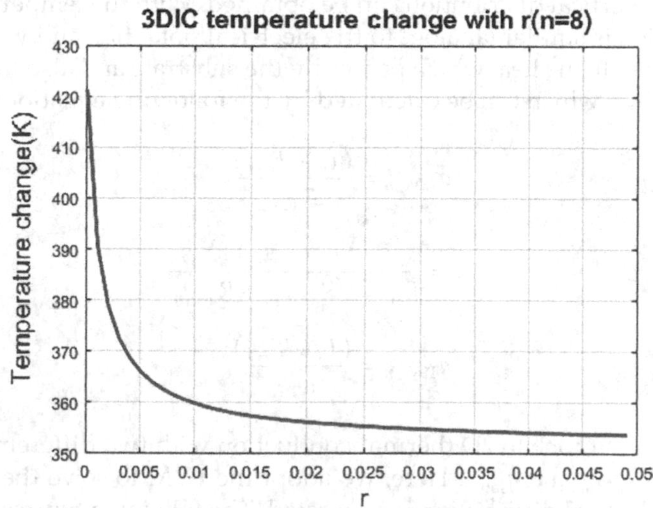

FIGURE 4.23
Temperature change with surface ratio r ($n = 8$).

where T is the temperature (K), k is the thermal conductivity (W/m/K), ρ is the density of the material (kg/m^3), c_p is the specific heat (J/kg/K), g is the power density of the heat sources (W/m^3), and h_i is the heat transfer coefficient in the direction of heat flow $\vec{i}$ on the boundary surface of the chip (W/m^2/K).

Figure 4.24 shows our total 3D thermal simulation work flow. In our case, we firstly get the surface ratio r, and calculate the r-dependent thermal resistance, such as R_{TSV}, R_{si}, R_{glue}, R_{ins}. By using the analytical vertical model, the

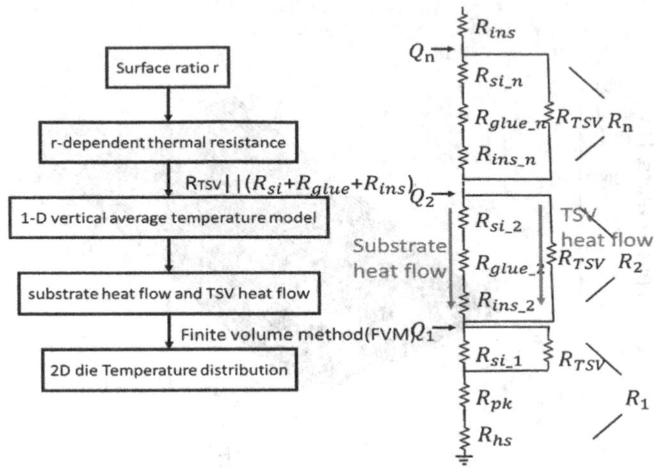

FIGURE 4.24
3D Thermal simulation work flow.

temperature vertical distribution can be obtained. With the temperature difference, which is similar (analog) to the electrical potential, and we can also directly calculate the heat which passes by the substrate and also heat which passes by TSVs, which can be calculated by the following equation:

$$q_{\text{TSV}\_n} = \frac{(T_n - T_{n-1})}{R_{\text{TSV}}} \tag{4.74}$$

$$q_{\text{sub}\_n} = \frac{(T_n - T_{n-1})}{R_{\text{si}\_n} + R_{\text{glue}\_n} + R_{\text{ins}\_n}} \tag{4.75}$$

$$q_{\text{TSV}\_n} = \frac{(T_n - T_{n-1})}{R_{\text{TSV}}}$$

Finally, we come back to 2D thermal conduction with two different kinds of heat source, q_{TSV_n} and q_{sub_n}. Here, we adopt the FVM to solve the problem. Then, a 2D thermal distribution is generated. The following figures show the bottom layer's and top layer's thermal temperature distribution in the substrate of an 8 dies model with $r = 0.01$ and 4 TSVs. The bottom layer's TSV temperature is 344 K and top layer's TSV temperature is 359.88 K (Figure 4.25).

- ***CMP (Chip multiprocessor) heat transfer***
  With the development of micro-processors, the traditional superscalar technology has slightly improved the CPU performance. Power dissipation issues have greatly restricted the gain of CPU clock frequency. Because of all these things, it needs a change of computer structure arrangement, and it needs a smaller distribution design of

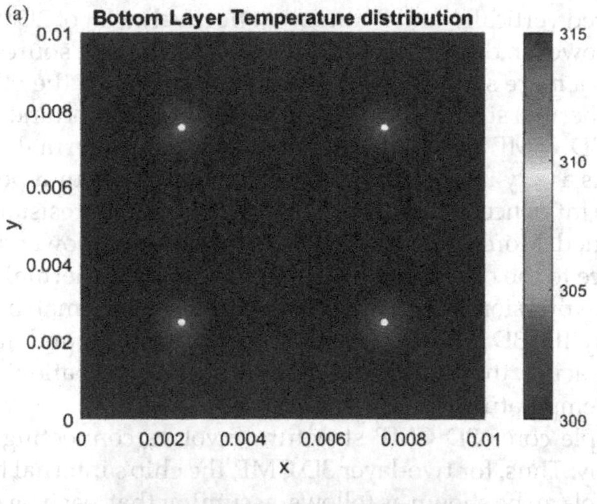

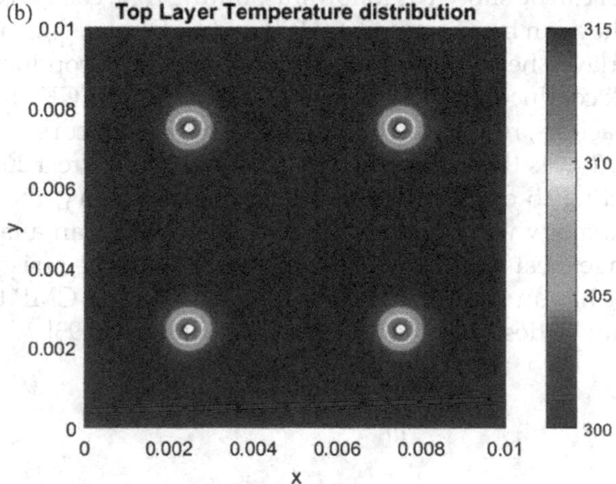

FIGURE 4.25
(a) Bottom and (b) top layer temperature distribution of 4 TSVs with surface ratio $r = 0.01$

micro-processor, and a partially better cell structure. In this background, researchers proposed Chip multiprocessor technology (CMP) [2]. CMP means to integrate two or more CPU cores in one chip, in which every core can do a multi-thread calculation simultaneously. The main purpose of CMP is to gain the integrity and increase performance. Nevertheless, with the augmentation of integration level and the reduction of feature size, chip internal interconnection line ratio increases continuously, especially when the chip feature size is reduced below 0.18 μm, the interconnection delay has surpassed gate delay; this brings a serious challenge for the CMP's performance. By adopting 3D technologies, the different cores can be connected vertically, and it can solve effectively a lot of 2D CMP problems. However, due to the 3D CMP layers multi-heat source, CMP 3D IC faces a more serious heat dissipation issues than the other 3D IC. So the thermal study of the CMP 3D IC is very important.

For 3D CMP temperature characteristics, thermal resistance matrix is a very important parameter. It represents unit power dissipation's influence to temperature, once the thermal resistance matrix is obtained. Moreover we can get whichever core's power dissipation influence to the other core. This chapter gives the thermal resistance matrix expression. Based on the thermal resistance matrix, we study precisely the 3D CMP temperature characteristics, and analyze the heat capacitor, thermal resistance and power dissipation's influence on the temperature.

Multiple cores 3D CMP structure involves connecting the cores vertically. Thus, for two-layer 3D CMP, the chip's internal heat transfer model can be shown as follows, assuming that, each layer has two cores. They can be expressed as *A*, *B* and *C*, *D*, g_{inter}, g_{intra} and g_{hs} are the interlayer heat conductance, inner layer heat conductance and heat sink conductance, respectively. CH, CI, CJ, and CK are the cores heat capacitors, respectively. PH, PI, PJ, and PK are cores' power dissipation. T_{amb} is the environment temperature (Figure 4.26).

According to current Fourier heat analysis theory, the heat flow has an analogy with the current, temperature has an analogy with the voltage, heat resistance has an analogy with the resistance, heat capacitor has an analogy with capacitance, and the CMP two layers model can be described as the following equation: [93]

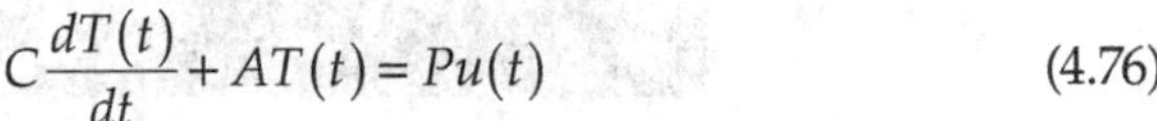

$$C\frac{dT(t)}{dt} + AT(t) = Pu(t) \quad (4.76)$$

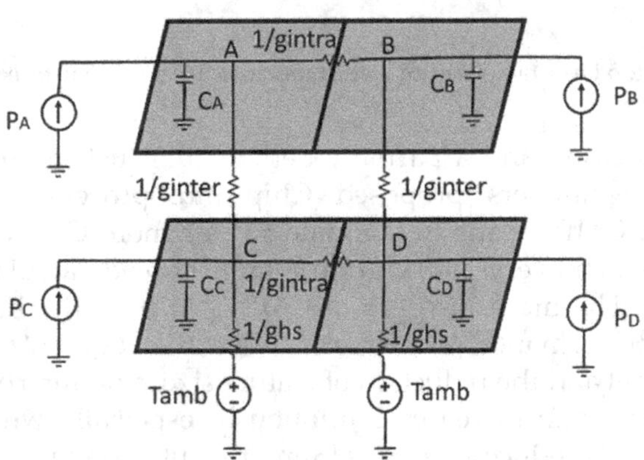

FIGURE 4.26
Four cores CMP model.

Here, C is a 4×4 diagonal matrix, and every diagonal element is the core's heat capacitor. $T(t)$ is a dimensional vector, which represents the four cores temperature at time t. A is a 4×4 thermal conductance matrix. P is a 4D power vector. $u(t)$ is a step function.

- ***Thermal resistance matrix***
  For the stationary case, we neglect the heat capacitor matrix C's influence, and the heat conduction equation can be simplified to:

$$T = A^{(-1)}P \tag{4.77}$$

Here, A^{-1} is the thermal resistance matrix. It describes unit power dissipation's influence to the temperature, so it can be called thermal influence matrix [3,93].

By using the nodal analysis, we use R_{ij} to represent the core i's influence resistance to core j, and the thermal resistance matrix can be expressed as follows:

$$A^{-1} = \begin{pmatrix} R_{AA} & R_{AB} & R_{AC} & R_{AD} \\ R_{BA} & R_{BB} & R_{BC} & R_{BD} \\ R_{CA} & R_{CB} & R_{CC} & R_{CD} \\ R_{DA} & R_{DB} & R_{DC} & R_{DD} \end{pmatrix} \tag{4.78}$$

Because core A and core D do not have a direct thermal connection, the thermal influence between them can be neglected, so as the B, C. Then:

$$R_{AD} = R_{DA} = R_{BC} = R_{CB} = 0 \tag{4.79}$$

By using the nodal analysis, the thermal resistance matrix can be expressed as follows:

$$A = \begin{pmatrix} g_{\text{intra}} + g_{\text{inter}} & -g_{\text{intra}} & -g_{\text{inter}} & 0 \\ -g_{\text{intra}} & g_{\text{intra}} + g_{\text{inter}} & 0 & -g_{\text{inter}} \\ -g_{\text{inter}} & 0 & g_{\text{intra}} + g_{\text{inter}} + g_{\text{hs}} & -g_{\text{intra}} \\ 0 & -g_{\text{inter}} & -g_{\text{intra}} & g_{\text{intra}} + g_{\text{inter}} + g_{\text{hs}} \end{pmatrix} \tag{4.80}$$

- ***CMP thermal modeling result***
  After getting the thermal resistance matrix, we can address the 3D CMP temperature characteristic equations:

$$\begin{pmatrix} C_A & 0 & 0 & 0 \\ 0 & C_B & 0 & 0 \\ 0 & 0 & C_C & 0 \\ 0 & 0 & 0 & C_D \end{pmatrix} \begin{pmatrix} \dfrac{dT_A(t)}{dt} \\ \dfrac{dT_B(t)}{dt} \\ \dfrac{dT_C(t)}{dt} \\ \dfrac{dT_D(t)}{dt} \end{pmatrix}$$

$$+ \begin{pmatrix} g_{\text{intra}} + g_{\text{inter}} & -g_{\text{intra}} & -g_{\text{inter}} & 0 \\ -g_{\text{intra}} & g_{\text{intra}} + g_{\text{inter}} & 0 & -g_{\text{inter}} \\ -g_{\text{inter}} & 0 & g_{\text{intra}} + g_{\text{inter}} + g_{\text{hs}} & -g_{\text{intra}} \\ 0 & -g_{\text{inter}} & -g_{\text{intra}} & g_{\text{intra}} + g_{\text{inter}} + g_{\text{hs}} \end{pmatrix}$$

$$\cdot \begin{pmatrix} T_A(t) \\ T_B(t) \\ T_C(t) \\ T_D(t) \end{pmatrix} = \begin{pmatrix} P_A \\ P_B \\ P_C \\ P_D \end{pmatrix} u(t) \tag{4.81}$$

Assume that every core has the same heat capacitor C, and their power dissipations are all P. By using MATLAB differential equation solver, we can get 3D CMP temperature characteristics (Table 4.2).

when $t = 0$, input a step heat source, 3D CMP temperature transient characteristics can be found below. Because core A and B stay in the same condition, so their temperature characteristics are the same. We can see that from the following figure and so is core C and D. From Figures 4.27 and 4.28, the change of the heat flow will increase the cores temperature at first. Then, the temperature will

TABLE 4.2

Parameters

P (W)	C (J/K)	g_{intra} (W/K)	g_{inter} (W/K)	g_{ins} (W/K)
25	5	0.41	6.67	0.82

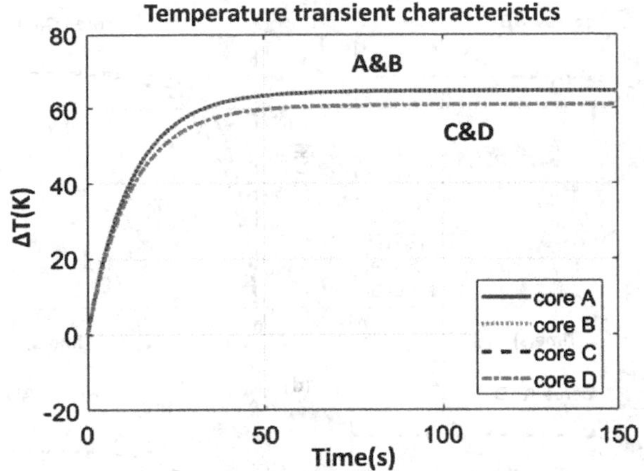

FIGURE 4.27
3D CMP transient temperature characteristics.

get stationary. Core *A* and *B*'s temperature is about 5K higher than that of core *C* and *D*, that's due to core *C* and *D* are a bit far away from the heat sink. Moreover, heat is not easy to be dissipated.

4.3.2.3 Electrothermal (ET) Modeling of VLSI Circuits

Noise coupling is a key issue in ICs [2,3,4]. It is aggravated in 3D circuits since multiple layers are placed above each other, allowing signals to propagate throughout the 3D structure. TSVs connect the different layers within the 3D system, carrying power, clock, and data. TSVs, seminal components of 3D technology, are short vertical interconnections between the different layers that can support global signaling requirements [1]. The TSVs penetrate the substrate of a layer and connect to either the first or last metal within that layer. Similar to two-dimensional (2D) substrate coupling, the signals within the TSVs can couple capacitive and inductive noises into the substrate, affecting closed victim circuits or other signals within or around a TSV.

A critical part of building a successful 3D system lies in the ability to physically arrange the circuitry, considering thermal management, wire length optimization, and communication overheads. The physical arrangement of logic blocks plays a crucial role in exploiting the advantages of short interconnects in 3D technologies and in addressing the disadvantages of thermal hot spots.

As indicated in the roadmap for ICs, the increasing power density needs the decreasing power supply voltage, which results in the need of larger current supply by the power delivery networks (PDNs) [5,20]. Consequently, Joule heating and power switching rise a lot the circuit temperature, which heavily influences the electrical performances, because of the heavily changed

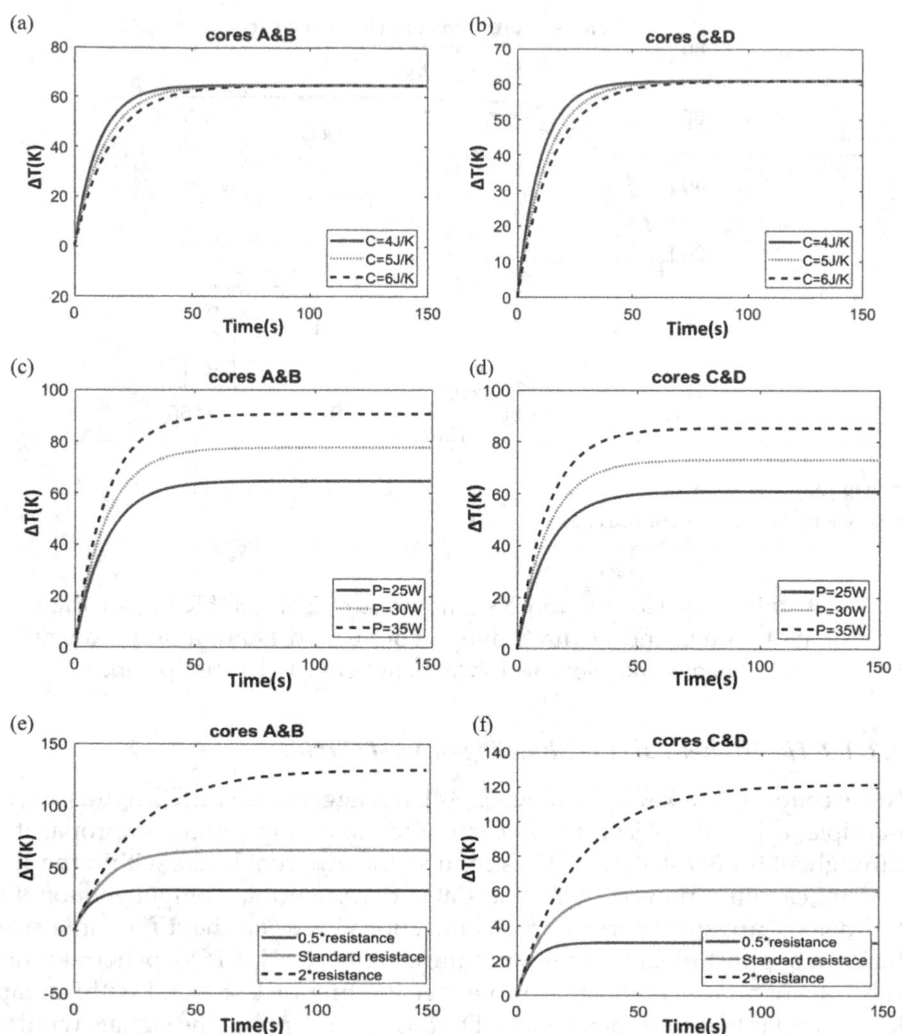

FIGURE 4.28
Temperature transient characteristics (a) of cores A & B with different heat capacities (b) of cores C & D with different heat capacities (c) of cores A & B with different power inputs (d) of cores C & D with different power inputs (e) of cores A & B with different thermal resistances (f) of cores C & D with different thermal resistances.

electrical temperature-dependent resistivity. So an efficient ET co-simulation PDN software is needed, for now, to calculate the effective IR drop, which is strongly related to the electrical resistivity. Moreover, ET co-simulations are inevitable in 3D IC design [69].

We use a standard structure for ground and power networks as shown in Figure 4.29. The ground and power networks are separated by an insulation

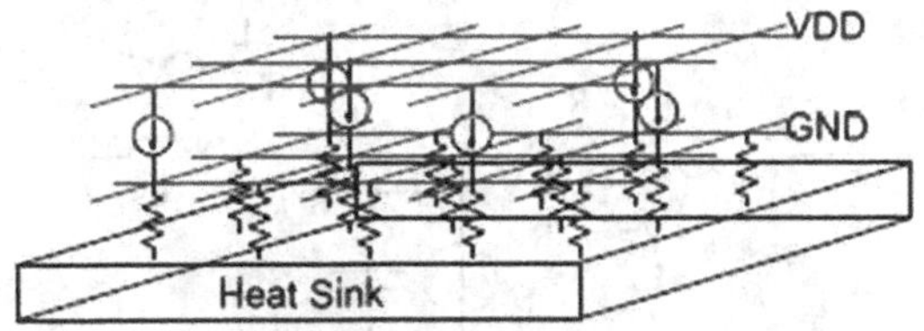

FIGURE 4.29
Structure for ground and power networks.

layer, and connected, respectively, with VDD and GND pins, through a series of impedance Z_{supply}.

The electrical and thermal issues should be solved together in the static and dynamic analysis. A device level thermal and electrical modeling method will be provided and justified, dedicated to CPW and TSVs. The main heat sources are the logic gates switching, as well as the Joule heating in the PDN interconnects. The thermal transport and electrical transport are very similar, the heat flow corresponding to the electrical current and the temperature difference plays the role of the electrical voltage. By this way, the electrothermal co-simulation can be exploited within the same circuit solver [93].

- ***Electrical modeling of VLS (very large scale) circuits***
  The DC problem is modeled by two regular resistive grids and one internal node; the temperature-dependent resistors $R(T)$ connect neighboring nodes of the same layer. The power and ground layer are interconnected by a current source I_0. The solution of the electrical problem can be expressed as the voltage drop at the node i:

$$V_d(i) = V_{DD} - \left(V_n(i) - V_g(i)\right) \cdot V_d(i) = V_{DD} - \left(V_n(i) - V_g(i)\right) \tag{4.82}$$

where $V_n(i)$ and $V_g(i)$ are the node potentials of power and ground planes, respectively.

A dynamic electrical model can also be carried out when parasitic dynamic circuit has been defined for this structure. We use the standard first order model for each physical cell, by considering the parasitic inductive effect L between adjacent nodes, and a parallel capacitive C between neighboring nodes on the VDD and GND grids. By this way, we draw a dynamic elementary cell schema as shown in Figure 4.30.

Due to the negligible heat diffusion in the dielectric, the thermal model can be assumed steady. So the temperature distribution $T(x)$ of metal interconnects can be solved by the 1D heat diffusion equation:

$$\frac{d^2T(x)}{dx^2} = -\frac{g(x,T)}{k_m} \tag{4.83}$$

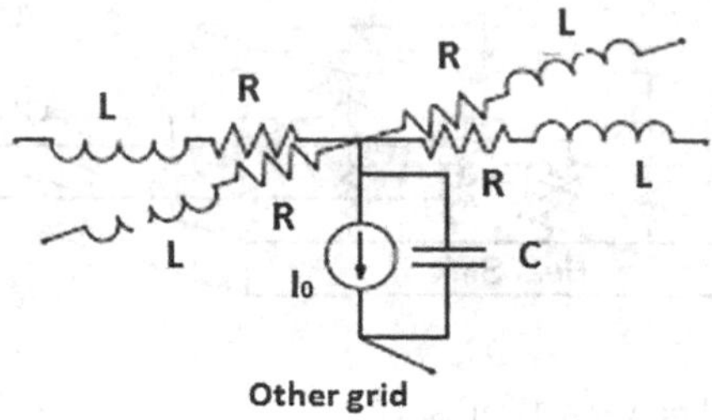

FIGURE 4.30
Dynamic electrical elementary cell thermal modeling of VLSi circuits.

where k_m is the thermal conductivity of the metal (W/mK) and $g(x,T)$ is the heat sources term (W/m^3).

Once the defined grid lengths are smaller than the heat characteristic length, the grid can be used to solve both electrical and thermal dissipation [93]. Figure 4.31 is a typical node of thermal network. It is composed of thermal resistances between neighboring nodes, and another thermal resistance linked to the heat sink. The VDD layers and GND layers are connected by adding the heat sources P_s and P_J; P_s is the switching power, defined as $P_s(i) = I_0 V_d(i)$, related to the voltage drop at node "*i*" calculated by electrical equation and P_J is the Joule heating term coming from the power dissipation in the PDN: [97]

$$P_J(i) = \frac{1}{2}\sum_{k=1}^{4} \frac{v_{ik}^2}{R_{ik}} \tag{4.84}$$

where k is the four neighboring nodes' index.

The electrothermal coupling problem is solved in a classical approach: firstly, solving the thermal network at room temperature, secondly calculating the temperature-dependent electrical resistances $R(T)$, thirdly solving the electrical network, and then getting the voltage. After calculating the thermal heat sources, we can repeat

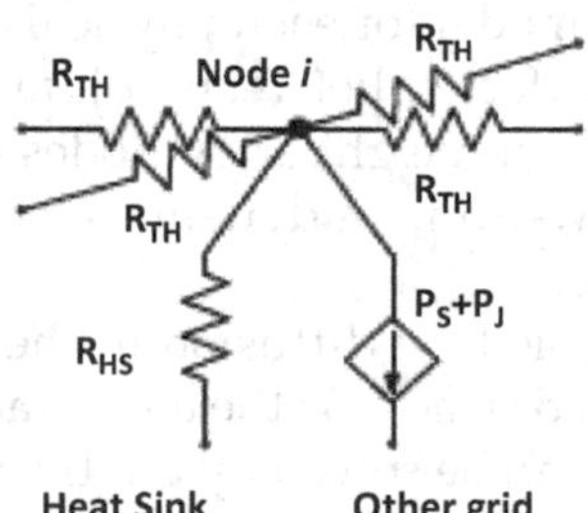

FIGURE 4.31
Dynamic thermal elementary cell.

the first three steps, and so on. The final temperature and voltage distributions can be obtained when the relative error gets smaller than the predefined error value.

Further investigation of the dynamic analysis can be carried out from the previous static ET analysis by adding the following two steps. The first step is calculating the dynamic elements (capacitors, inductors, etc.…) related voltage drop. The second step is checking if the new thermal distribution is still compatible with the static electrothermal equilibrium. Thus, with a predefined impedances circuit, a frequency analysis can be then carried out by using the dynamic analysis workflow.

3D integration gives a way to higher performance, increased functionality and heterogeneous implementation. Nevertheless, compared to 2D integration, due to the parasitic, large switching current and longer power delivery path, the power and thermal integrity issues are always the concerns for 3D integration. In this part, we provide a study of power and thermal issues of 3D integration, especially for the TSVs. Our study is based on a three layer's system using TSV based technologies. Based on the heat dissipation models, we present a power and thermal integrity analysis of 3D ICs, especially TSV.

- ***Pulse signal definition***
  Our study takes the dynamics power analysis into consideration by combining parasitic of the 3D multi-layers system. The extracted parasitic deals with the coplanar line, redistribution layer and TSVs. Coplanar line parasitic are modeled by resistance, capacitance and inductance. Redistribution layer is modeled by resistance, inductance, and capacitance. The time-dependent voltage sources can be represented by the electrical characteristics: (1) switching frequency; (2) rise and fall times; (3) duty cycle; (4) delay time and (5) peak voltage. Here, in our study, the switching circuits are directly replaced by rectangular waveforms;

Switching pulse-like voltage sources are expressed as follows:

$$V(t)=\begin{cases} V_{\min},\ 0<t<t_{\text{delay}} \\ \left(t-t_{\text{delay}}\right)*\dfrac{V_{\text{peak}}-V_{\min}}{t_{\text{rise}}},\ t_{\text{delay}}\le t<t_{\text{delay}}+t_{\text{rise}} \\ V_{\text{peak}},\ t_{\text{delay}}+t_{\text{rise}}\le t<t_{\text{delay}}+t_{\text{rise}}+T*d \\ V_{\text{peak}}-\left(t-t_{\text{delay}}-t_{\text{rise}}-T*d\right)*\dfrac{V_{\text{peak}}-V_{\min}}{t_{\text{fall}}},\ t_{\text{delay}} \\ \qquad +t_{\text{rise}}+T*d\le t<t_{\text{delay}}+t_{\text{rise}}+T*d+t_{\text{fall}} \\ V_{\min},\ t>t_{\text{delay}}+t_{\text{rise}}+T*d+t_{\text{fall}} \end{cases} \tag{4.85}$$

The analysis is established on a 3D multi-layers structure. Models are constructed from 3D-TLE extractor which can extract and generate the SPICE compatible interconnects' model. The key challenge of this analysis is to calculate simultaneously electrical and thermal characteristics. Therefore, the temperature-dependent resistivity is used in this analysis.

$$\rho = \rho_0\left[1+\beta(T-T_0)\right] \tag{4.86}$$

where ρ_0 ($1.53265e^{-8}$ Ω·m) is the electrical resistivity at T_0 which is 300K and β (0.0067) is the temperature coefficient for electrical resistivity [93]. The electrical resistivity of conductors goes up while temperature increases, and the power supply will be greatly influenced. Besides, the voltage drop and flowing current on the transmission line will cause Joule heating. After a sufficiently long time, the difference of temperature distribution will lead to EM issues. By this method, electrical, and thermal issues are coupled together.

To perform the 3D multi-layers system for electrical analysis, we adopt Kirchhoff's law to solve the linear equations. These linear equations can be transformed together to one matrix. The matrix formulation follows the rules of modified nodal analysis (MNA) method which is often used for power grid analysis [69,93]. These matrices can be expressed as Laplace transforms:

$$[G+sC]x(s)=I(s) \tag{4.87}$$

where G is the $n \times n$ conductance matrix, C is the $n \times n$ capacitance and inductance matrix, and I is the $n \times 1$ vector of current sources and voltage sources. $x(s)$ is the $n \times 1$ vector that represents nodes' voltage and current. To get nodes' voltage drop and current distributions, we solve the electrical power network by inversing this matrix $[G+sC]$. For large system analysis, we use sparse matrix and generalized conjugate residual (GCR) method to ease matrix computations.

As a heat conductor from one layer to another, TSVs can sometimes be very beneficial for dissipating heat away from a hot layer but sometimes they can be harmful to neighboring layers or components. In this work, by following the principle of electrical-thermal duality [93], the heat flow passing through a thermal resistor is equivalent to the current passing through the electrical resistance and temperature difference is equivalent to voltage difference.

Heat is normally generated from the upper layers, and transferred through TSVs and other layers (for example, isolation layers) to the silicon substrate. Isolation layers have much lower thermal conductivities than that of silicon; so they act as thermal barriers. The voltage changes in the switching circuits will contribute to the heat generation. TSVs, redistribution layers, coplanar lines and heat sink are modeled as thermal resistances in order to conduct the vertical and lateral heat flows in the 3D system. Due to the difference between TSVs' vertical thermal resistances and other material layers' thermal resistances, heat dissipates at different rates. The equivalent thermal resistances are represented as the parallel connection of TSVs' resistances and other layers' thermal resistances. The temperature of heat sink is assumed to be the ambient temperature. Thanks to the principle of electrical-thermal duality, the thermal network can also be solved by using MNA method. The nodes' thermal linear equations can be merged to the following equation:

$$G_{th}T = Q \tag{4.88}$$

where G_{th} is the thermal resistance matrix of size $m \times m$ and m is the number of nodes on the 3D system. T is the $m \times 1$ vector of temperature nodes and Q is the $m \times 1$ vector for heat sources.

We develop an electrical and thermal analysis flow to calculate the electrothermal coupling effect of a 3D system. The proposed method is an iterative process; Figure 4.32 shows the electrical and thermal analysis flow.

Electrical analysis is a transient one; the voltage drop on the metal tracks will result in an elevation of temperatures. The increase in temperature is then treated as a thermal source and applied to the thermal analysis step. Thermal analysis provides temperature distribution for the different layers and TSVs, where altered temperatures will change conductors' resistivity and then influence electrical analysis. The iterative calculation of resistivity and temperature will

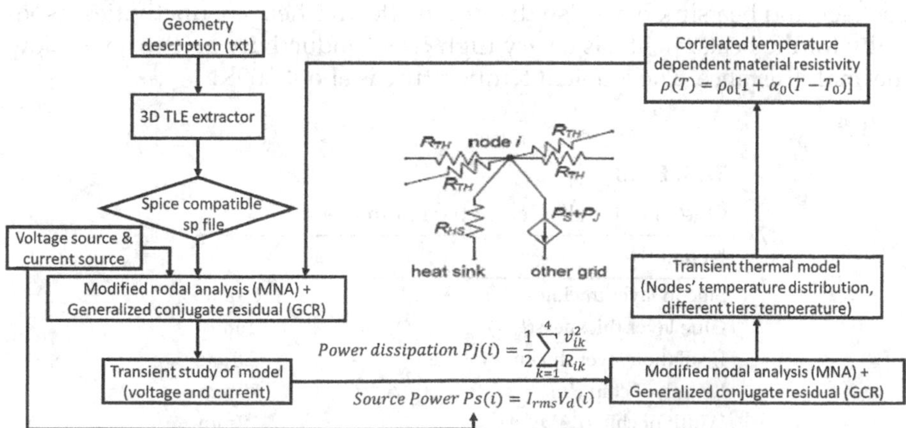

FIGURE 4.32
3D IC electrothermal coupling flow.

converge when there are no more changes on voltage drop and temperature distribution. Here, the thermal analysis is a steady-state analysis which captures the temperature distribution from a given electrical network. This method is limited to capture the dynamic thermal distribution. However, as fast and accurate capturing of temperature distribution is always our purpose. The 1D static thermal analysis is a viable approach. We adopt the MNA as described above to perform efficient electrical and thermal networks analysis.

4.3.2.3.1 Case I: One via Model

The one via model is shown in Figure 4.32 (see Table 4.3). The chip is composed of one coating layer, one oxide layer, and one silicon layer; the center is a copper TSV. The bottom is attached to a package and outside the package, a heatsink is used. The chip's size is 10 mm × 10 mm, the silicon layer's thickness is 50 µm, and the coating layer and insulation layer have, respectively, a thickness of 2 and 8.8 µm. TSV's surface ratio r is fixed to 0.01.

The electrical circuit and thermal circuit are shown in Figure 4.33a,c. We input a pulse signal (defined as previous part), and parallel a 50 Ω resistance at the input and output, and a 150fF capacitance at output in order to see the output signal. Here, the resistance of TSV is about $0.0052e^{-4}$ Ω; so as compared with 50 Ω, the TSV's resistance is very small. So the parallel circuit's effective resistance depends on the TSV resistance (frequency-dependent straight wire formula and the parallel connection of 50 Ω resistance will not change the output signal. The current will induce heat when passing by the resistance of TSV. The generated heat is dissipated in the 3D geometry. The thermal path is composed of TSV conduction and the other three layers' serial heat conduction, the heat pass separately by the series of three layers and also by the TSV (because of the high copper's heat conductivity). Then the heat is collected in the package level, and conducted through the package to the heat sink. The package and heatsink have also their own effective heat conductivities; especially for the heatsink, it has a very high-heat conductivity and so a very low thermal resistance. The ambient temperature is about 300K.

TABLE 4.3

One via Model's Geometry Parameters

Geometry	
Silicon layer thickness(l_{si})	50 µm
Glue layer thickness(l_{glue})	2 µm
Insulation layer thickness (lines)	8.8 µm
Length of chip (L_x)	10 mm
Width of chip (L_y)	10 mm
TSV number	1
TSV ratio (r)	0.01

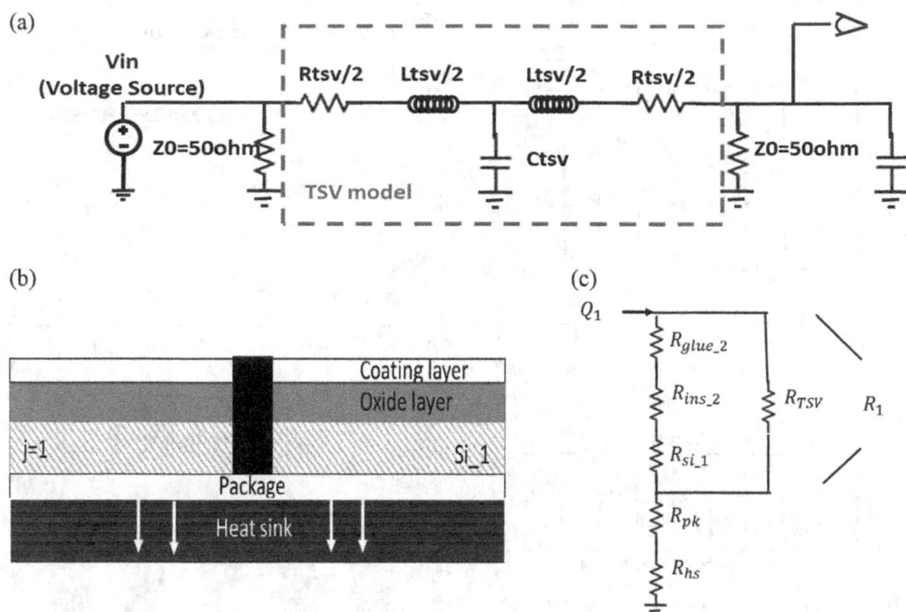

FIGURE 4.33
(a) Electrical compact mode. (b) 2D Geometry. (c) Thermal compact model of one via model.

As shown is Figure 4.34, the *S*-parameter is calculated by using ADS simulator.

Figure 4.35 shows the 2.5 GHz's signal transmission effects. Figure 4.35a shows the comparison of input and output signal. Figure 4.35b shows, respectively, the top of coating layer, the bottom of TSV and the bottom side of packaging's temperature. Figure 4.35c shows the voltage transmission change

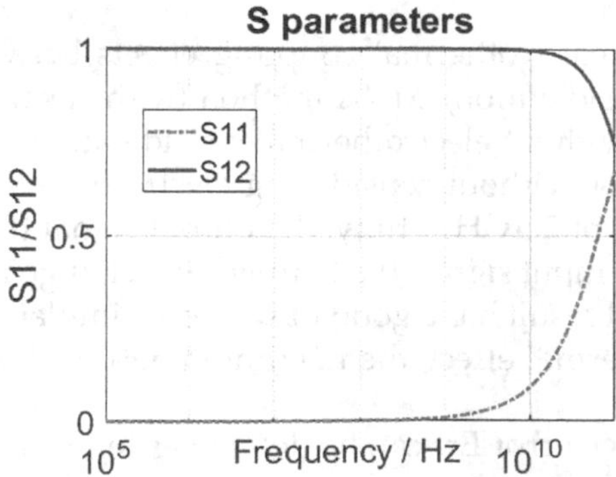

FIGURE 4.34
S-parameters of 1-TSV model from ADS (valid for frequency below 20 GHz and of quasi-static assumption).

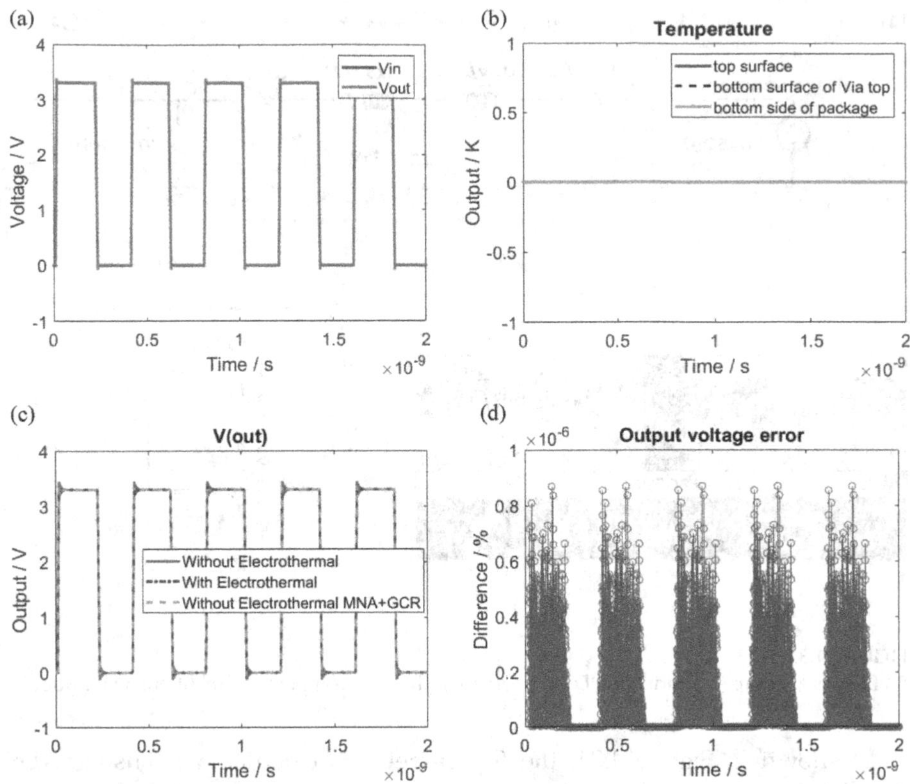

FIGURE 4.35
Signal transient integrity analysis for 1-TSV model 2.5GHz (a) Input vs Output signal. (b) Maximum temperature rise. (c) Comparison between ADS (without electrothermal consideration), MNA method (with electrothermal consideration) and MNA (without electrothermal consideration). (d) Output voltage error.

when considering electrothermal coupling effects between ADS (without electrothermal consideration), MNA method (with electrothermal consideration) and MNA (without electrothermal consideration). Figure 4.35d is the output voltage error without considering electrothermal effect. As we can see from the result of 2.5GHz study, the output signal has almost the same waveform than the input signal, the temperature change in the circuit is fine. The ADS simulated result has a good fit with our simulator, and after considering the electrothermal effect, the maximum output change is 0.2%, which can be neglected.

The output voltage error Err can be defined as follows:

$$\text{Err} = \frac{\max\left(\text{abs}\left(V_1 - V_2\right)\right)}{V_{\max\_in}} \tag{4.89}$$

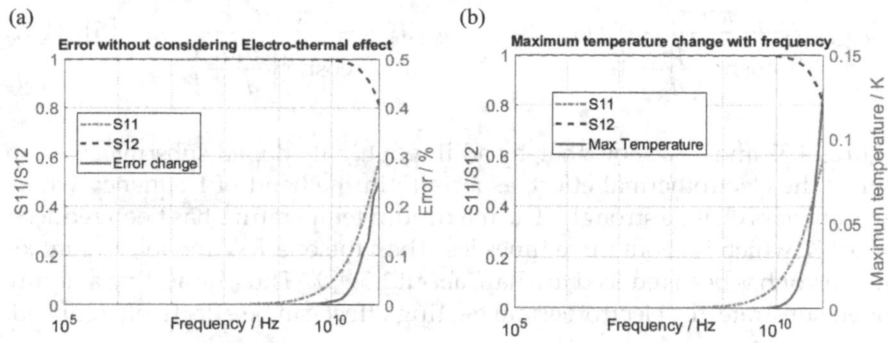

FIGURE 4.36
S-parameters with (a) error curve and (b) maximum temperature curve for one via model (valid for frequency below 20GHz and of quasi-static assumption).

where V_{max_in} is the input maximum voltage, V_1 is the first transient voltage, and V_2 is the second transient voltage.

From the above results, the one TSV model does not have electrothermal effect at very low working frequency. For some period of frequencies, the output frequency increases linearly, and it will generate a lot of heat and increase the via's temperature. Therefore, we find out that the electrothermal effect has a relationship with the trend of *S*-parameters. The *S*-parameters stay stable at low frequency and very high frequency, but for a band of frequency, it changes dramatically. We can superpose the *S*-parameters with the error curves and maximum temperature curve. Figure 4.36 shows the result. But, due to the non-linearity system, from 20GHz to 10 THz, the results should use full-wave models for electrical study. The current result is no more valid in this band of frequencies.

As expected, the *S*-parameters have a good tendency versus the electrothermal effects. Strong electrothermal effects appear while *S*-parameters change.

4.3.2.3.2 *Case II: One TSV Model with Substrate RC Effect*

The only difference between this model and the previous model is the substrate. In the previous model, the substrate is treated directly as a ground node. But in this model, the substrate is highly doped. It is treated as a conductor with a very low resistivity (1e^{-4} $\Omega \cdot$ m). The substrate is expressed as a RC parallel model. The substrate resistance and capacitance can be expressed by the following equations:

$$C_{S_i\_sub} = \frac{\pi \times \varepsilon_0 \varepsilon_{r,si}}{\cosh^{-1}\left(\frac{p_{TSV}}{d_{TSV}}\right)} \times h_{TSV}[F] \quad C_{S_i\_sub} = \frac{\pi \times \varepsilon_0 \varepsilon_{r,si}}{\cosh^{-1}\left(\frac{p_{TSV}}{d_{TSV}}\right)} \times h_{TSV}[F] \tag{4.90}$$

$$G_{S_i\_sub} = \frac{\pi \times \sigma_{si}}{\cosh^{-1}\left(\frac{p_{TSV}}{d_{TSV}}\right)} \times h_{TSV}[S] \qquad G_{S_i\_sub} = \frac{\pi \times \sigma_{si}}{\cosh^{-1}\left(\frac{p_{TSV}}{d_{TSV}}\right)} \times h_{TSV}[S] \quad (4.91)$$

Figures 4.37 and 4.38 show that, by adding a highly doped substrate, we can reduce the electrothermal effect, especially in the band of frequency where *S*-parameters change strongly. The maximum temperature has been reduced to 5e^{-4} K, which is about three times less than the one TSV model, and maximum error has been reduced to a half, about 1.5e^{-5}%. Thus, by adding a highly doped substrate, the electrothermal heating effect can be effectively reduced.

4.3.2.3.3 Case III: One TSV Model with a Very High Thermal Resistance

In order to get a higher temperature and see a more obvious difference of signal transmission, we set R_{hs} and R_{pk} in Figure 4.37c to 2,000 K/W, respectively.

From Figure 4.39, we can see that the error curve and maximum temperature curve have always the same tendency. They keep stable at low frequency. Once the *S*-parameters change, the two curves change dramatically, although we have performed a study on the electrothermal modeling until 10 THz; due to the non-linear Maxwell's equations system, from 20 GHz

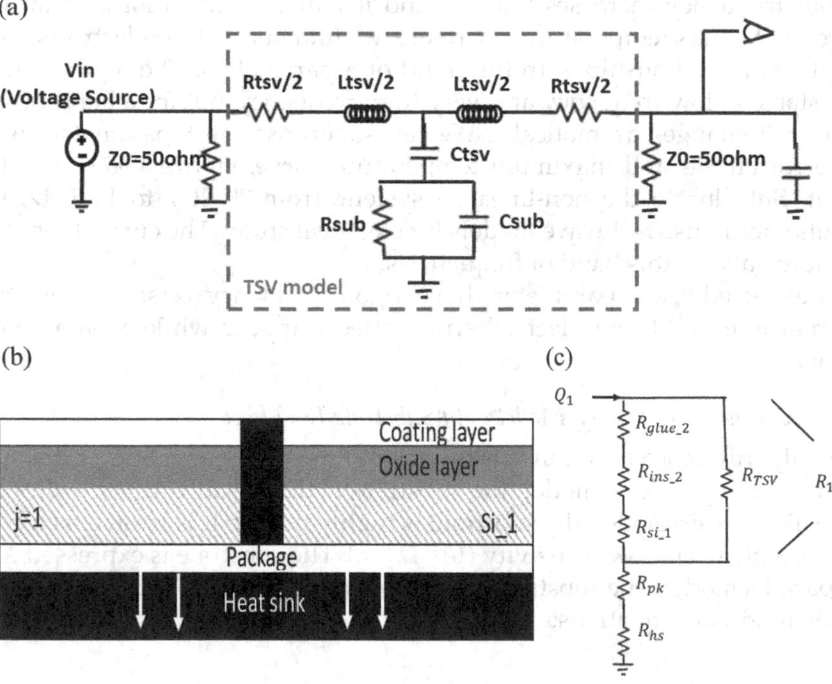

FIGURE 4.37
(a) Electrical compact mode. (b) 2D Geometry. (c) Thermal compact model of one via with substrate model.

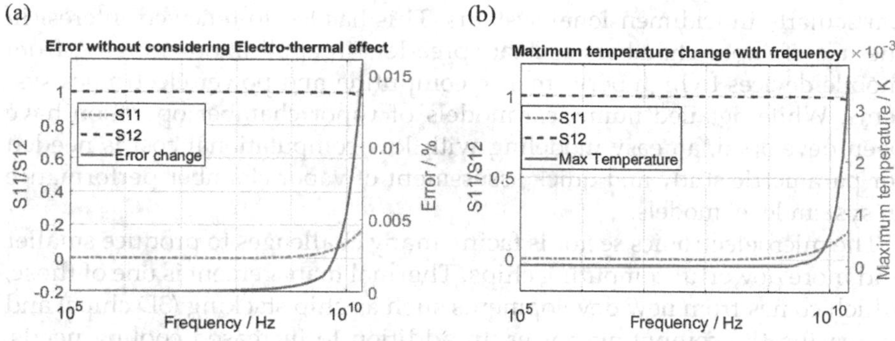

FIGURE 4.38
S-parameters with (a) error curve and (b) maximum temperature curve for one via with substrate model (valid for frequency below 20 GHz and of quasi-static assumption).

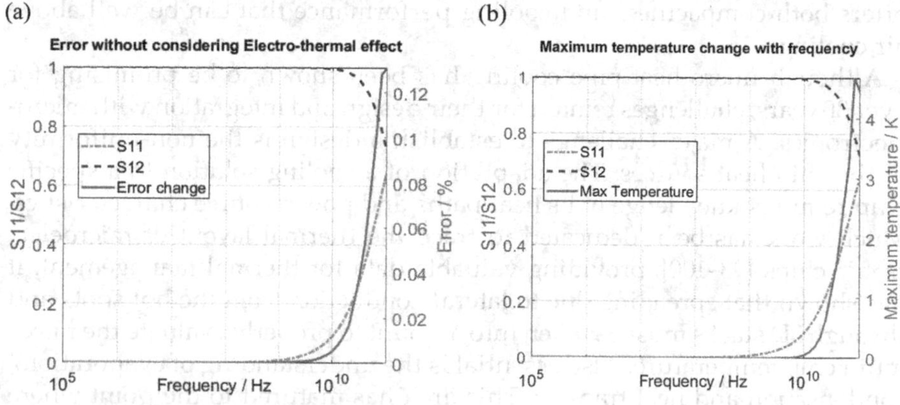

FIGURE 4.39
S-parameters with (a) error curve and (b) maximum temperature curve for one via with high thermal resistance model (valid for frequency below 20 GHz and of quasi-static assumption).

to 10 THz, we should use full-wave models for electrical study. But at least, it works for the frequencies less than 15 GHz, sufficient for our industrial purposes, for the next generation of mobile sample, for instance. Here, we will not speak on quantic interference of temperature, for nano devices. This analysis method will be useful for thermo-mechanical effects.

4.4 Heat Pipe

Advances in the computational performance of electronic devices have created a clear need for improved methods of passive thermal management, particularly in tridimensional systems. This has led to renewed interest in the use of vapor chambers as heat spreaders in applications ranging from mobile devices to high-performance computing and power electronics systems. While detailed numerical models of vapor chamber operation have been developed, an easy modeling with low computational cost is needed for parametric study and quick assessment of vapor chamber performance in system-level models.

The microelectronics sector is facing many challenges to produce smaller and more powerful computing chips. Thermal management is one of those, which comes from new developments such as chip stacking (3D chips) and increasing the computing power. In addition to increased cooling needs, the mobile device market tends to reduce the space available for heat dissipation. In this context, conventional air heat sinks are no longer adequate and alternative compact and high-heat flux cooling approaches are required. Micro-heat pipe cooling with porous medium is an approach that offers both compactness and cooling performance that can be well above air cooling.

Although micro-heat pipe cooling has been shown to be promising for over 30 years, challenges remain for their design and integration with microelectronics. A major challenge to establish a design is the non-uniformity of the chip heat sources. The adaptation of a cooling solution to a specific chip requires knowledge of its heat paths and power source characteristics. Much work has been dedicated to trace the thermal layout of microelectronic chips [98–100], providing valuable data for thermal management. It has shown that spreading due to lateral conduction from the hot spots and through 3D stacks must be taken into account to properly evaluate the maximum chip temperature. Also essential is the understanding of evaporation/condensation and heat transfer. This area has matured to the point where both static and transient analytical tools to design micro-heat pipe are now available [100,110].

However, it is essential to merge both the chip thermal conduction model and the micro-heat pipe cooling system together to take into account their coupled effects. This aspect is especially relevant when designing the cooling system to efficiently use the length-thickness ratio. Reducing the length-thickness ratio helps to gain the benefit from the heat sink, but it loses this benefit when is smaller than a certain ratio.

Relatively simple analytical models were proposed to define the chip thermal scheme layout [108,111]. Although convenient for uniform planar properties and heat source, they are rapidly limited when comes time to simulate the heat spreading on complex structures. Simulating heat spreading is critical as it generally has an important impact on the chip thermal resistances. Some authors, such as Amir Faghri et al. [105,107], gave the analytical approach of classical heat pipe, including single side heating and cooling heat pipe, double side heating and cooling, and multi-heat sources heating. However, such models do not include multiple materials in a single layer or a 3D complicated circuit. The complex geometry of true 3D IC products emphasizes the limitations of such analytical models.

Numerical models dedicated to 3D chip have therefore been developed. They generally represent the chip and micro-channels by a thermal resistance network and are more adapted to tackle the conjugate heat transfer in 3D chips. They showed good accuracy compared to full 3D finite-element (FEM) or experimental tests for a fraction of the computing time, which makes them a valuable asset for thermal study in early-design phases [111–113].

A similar 3D chip model is adopted in this work, but by combining commercial finite-element software to model conduction in the chip. In comparison with other similar models, the use of commercial software readily allows multiphysics studies, such as thermos-fluidic behavior due to micro-heat pipe cooling, as well as the use of extensive post-processing capabilities. This model is used to gain insight on the design parameters for micro-heat pipe embedded on a standalone 3D chip.

In this part, the heat transfer in the micro flat heat pipe (FHP) is first described and the different limits are calculated. Then, the modeling approach for a standalone 3D chip is presented. A parametric analysis is done on the standalone chip. Finally, relevance of the method is discussed.

4.4.1 2D Modeling Approach

- *Thermal resistances of FHP*

  Before detailing the analytical-numerical model structure, the relation between the thermal resistances is first studied to both understand the model itself and interpret the results. As shown in Figure 4.40, heat goes through different thermal resistances as it flows from the heat source to the cooling fluid and essentially corresponds to Garimella et al.'s [109] approach.

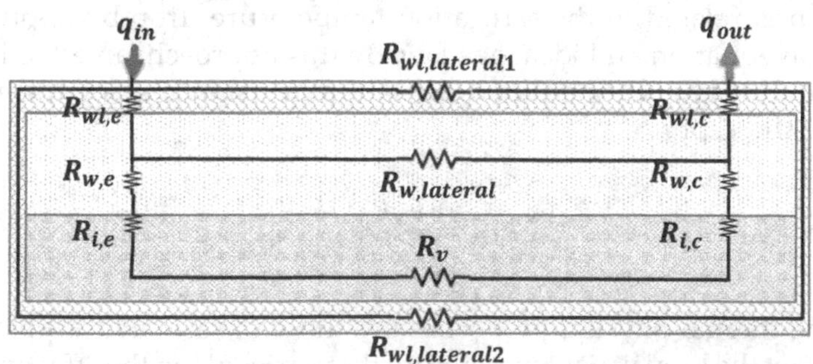

FIGURE 4.40
Schematic of approximate network of thermal resistance of FHP.

4.4.1.1 2D Heat Pipe Analytical Solution

The temperature drop due to the primary heat transport mechanisms that occur in each section of the heat pipe can be represented using the simplified effective thermal resistance network, as shown in figure. The thermal resistances due to conduction through the thickness of the wall in the evaporator ($R_{wl,e}$) and condenser ($R_{wl,c}$) sections are negligible (because of the thin thickness of wall), which is targeted at ultrathin heat pipes constructed from high-conductivity materials. Lateral conduction along the heat pipe wall is shown as $R_{wl,lateral1}$ and $R_{wl,lateral2}$; these thermal resistances can be calculated using an effective device length that considers the varying heat load along the length of the evaporator and condenser sections, and is given by

$$L_{eff} = L_a + \frac{L_e + L_c}{2} \tag{4.92}$$

Resistances through the wick in the evaporator ($R_{w,e}$), condenser ($R_{w,c}$), and laterally across the wick ($R_{wl,lateral}$) are assumed to be due to conduction only. Convective heat transfer in the porous medium is neglected due to the small interstitial liquid velocities; hence, the liquid flow field does not need not be solved. An effective saturated wick thermal conductivity is used to estimate these 1D conduction resistances. Lateral heat flow through the wick is neglected, owing to the high resistance that results from the low thermal conductivity and high aspect ratio wick geometry. To keep the model general, heat loss to the surroundings is neglected.

The interfacial phase-change thermal resistances at the wick–vapor interface in the evaporator and condenser sections, represented by $R_{i,e}$ and $R_{i,c}$, respectively, are typically small and, therefore, neglected. The effective thermal resistance of the vapor core R_v, is calculated by representing the vapor flow field as incompressible, laminar, fully developed flow between parallel plates, and governed by continuum physics, to estimate the pressure drop in the vapor core over the effective length. This pressure drop is related to the saturation temperature drop by applying the Clapeyron equation and ideal gas law. By this approach, an effective thermal conductivity is defined for a 1-D lateral resistance along the effective length of the vapor

$$k_{eff,v} = \frac{h_{fg}^2 P_v \rho_v t_v^2}{12 R \mu_v T_v^2} \tag{4.93}$$

where the thermo-physical properties are those of saturated vapor and are evaluated at the local vapor temperature that varies along the effective length of the device. To implement temperature-dependent vapor properties, the 1D vapor thermal resistance is discretized, and each individual resistance is back-calculated from the known condenser-side temperature, and by iterating upon the heat flux passing through the vapor core. This approach does not consider convection effects within the vapor core.

Combining the system resistances yields the effective heat pipe resistance:

$$R_{HP} = \frac{1}{\dfrac{1}{R_{wl,lateral1}} + \dfrac{1}{R_{wl,lateral2}} + \dfrac{1}{\left(R_v + R_{w,e} + R_{w,c}\right)}} \quad (4.94)$$

- ***Performance-limiting conditions***
  By comparing the performance across different input heat fluxes and for a range of geometries, it is possible to identify the limiting conditions that become predominant at small thickness and low input power operation. Figure 4.41 shows a contour map of the ratio between the heat pipe and copper heat spreader thermal

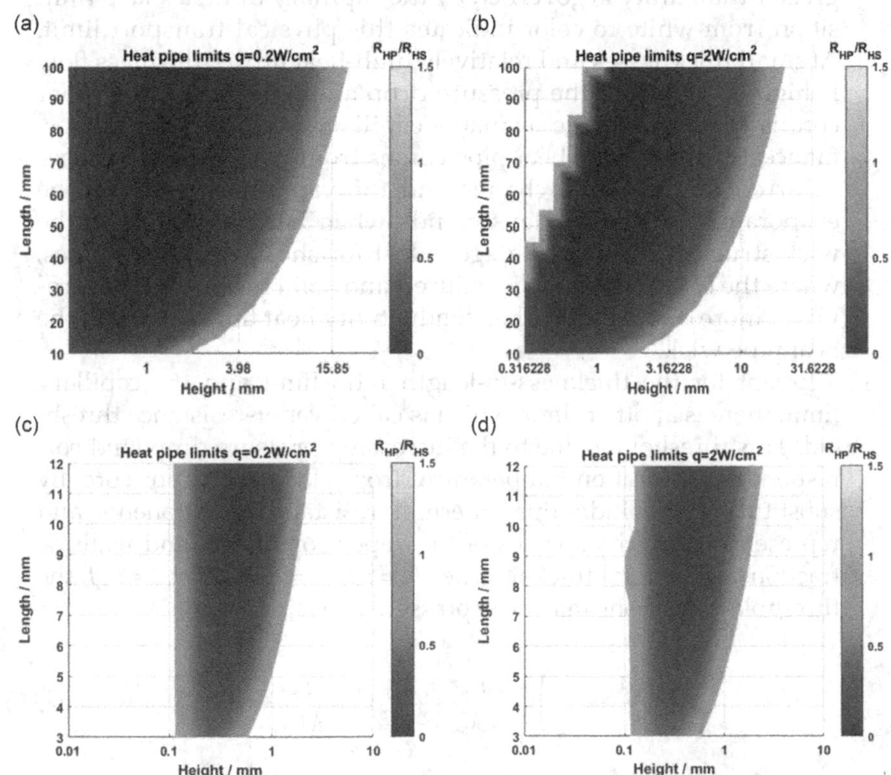

FIGURE 4.41
Contour map of the resistance ratio R_{HP}/R_{HS} plotted as a function of adiabatic length and thickness for an input heat flux of (a) 0.2 and (b) 2 W/cm² for length from 10 to 100 mm, of (c) 0.2 and (d) 2 W/cm² for length from 3 to 12 mm.

resistances with the adiabatic length and total thickness being varied. The colored area indicates an advantageous heat pipe performance and spreads to white as the ratio changes to the copper heat spreader. The performance thresholds and the capillary limit are also indicated. A pressure of 2,250 Pa is set as the maximum capillary pressure.

Figure 4.41a,c shows the limiting conditions for the case of a much lower input heat flux of 0.2 W/cm^2. At low input heat fluxes, there is sufficient capillary pressure due to the reduced fluid velocities, and the capillary limit dramatically shifts to lower thicknesses and higher working lengths.

Figure 4.41b,d shows the resistance ratio contour plot for a relatively high input heat flux of 2 W/cm^2 to illustrate the limits encountered when designing ultrathin heat pipes for moderate power dissipation. In this case, as the thickness is reduced at a given length, the minimum thickness for which R_{HP}/R_{HS} remains greater than unity is governed by the capillary limit; a sharp transition from white to color indicates this physical transport limit. At small thicknesses and relatively high-heat fluxes, the mass flow is high enough that the pressure drop along the wick and vapor core is higher than the available capillary pressure. The performance advantage of a heat pipe comes from the small resistances incurred due to phase change and the vapor core between the evaporator and the condenser ends when heat is applied to the wick structure. This advantage is lost for short, thick heat pipes, where the effective length is reduced and solid heat spreaders provide a more direct and higher conductivity heat flow path than the heat pipe wick.

Except for the thickness-to-length ratio limit and the capillary limit, there is another limit, which is called vapor-resistance threshold. This threshold is due to the increasing pressure drop (and corresponding saturation temperature drop) along the vapor core. By substituting the individual thermal resistance components and representing the thicknesses of the vapor core, wick, and walls as fractions of the total thickness, i.e., $t_v = r_v t$, $t_w = r_w t$, and $t_{wl} = r_{wl} t$, the threshold conditions may be expressed as [114]

$$\frac{r_w}{k_w}\left(\frac{1}{L_e}+\frac{1}{L_c}\right)t^4-\frac{L_{eff}}{k_s}\frac{1}{1-r_w}t^2+\frac{L_{eff}}{M_v r_v^3}=0 \tag{4.95}$$

Solving the equation results in two positive roots representing the thickness at both threshold conditions (vapor-resistance and high thickness-to-length ratio). The limiting minimum thickness governed by the vapor-resistance threshold is as follows:

$$t_{\text{limit}} = \left(\frac{\frac{L_{\text{eff}}}{k_s} a - \sqrt{\left(\frac{L_{\text{eff}}}{k_s}\right)^2 - 4\frac{r_w}{k_w}\left(\frac{1}{L_e} + \frac{1}{L_c}\right)\left(\frac{L_{\text{eff}}}{M_v r_v^3}\right)}}{2\frac{r_w}{k_w}\left(\frac{1}{L_e} + \frac{1}{L_c}\right)} \right)^{\frac{1}{2}} \quad (4.96)$$

where a = $1/(1 - r_{wl})$ and M_v is a constant representing the vapor properties. This limiting thickness is independent of the heat input; however, the capillary limit should be evaluated in conjunction with this threshold, which would be expected to prevail at any moderate heat inputs. The vapor properties dictating the vapor-resistance threshold are represented as a single factor that can be used as a merit number for fluid selection in the design of the ultrathin heat pipes operating at low heat inputs

$$M_v = \frac{h_{fg}^2 \rho_v P_v}{R \mu_v T_v^2} \quad (4.97)$$

which is independent of geometric parameters.

The classical configurations of the FHPs, which consist of the wall, the porous wicks saturated with working fluid, and the vapor space are shown in Figure 4.42a–c: the configurations for FHPs with a single heat source, multiple heat sources, a heat source at the bottom and a heat sink on the top of the heat pipe, and multiple heat sources with symmetric boundary conditions, respectively. The underlying assumptions are two-dimensional wall temperature and vapor flow, at steady-state, an incompressible and laminar flow, a saturated wick, constant material properties, constant saturation temperature, and linear temperature profile across the thin wick structure, are considered.

- ***Conduction in wall***

  The non-dimensional temperature, heat flux and coordinates are defined as follows:

$$\vartheta = \frac{k_w}{q_e h_w}(T - T_{sat}) \quad (4.98)$$

$$X = \frac{x}{l}; Y = \frac{y}{h_w}; H_w = \frac{h_w}{l}; H_l = \frac{h_l}{l}; L_e = \frac{l_e}{l}; L_c = \frac{l_c}{l}; \gamma = \frac{L_e}{L_c} \quad (4.98\text{bis})$$

  The 2D dimensionless steady-state heat conduction equation in the wall with constant thermal conductivity can be written as follows:

$$\frac{\partial^2 \theta}{\partial X^2} + \frac{1}{H_w^2}\frac{\partial^2 \theta}{\partial Y^2} = 0 \quad (4.99)$$

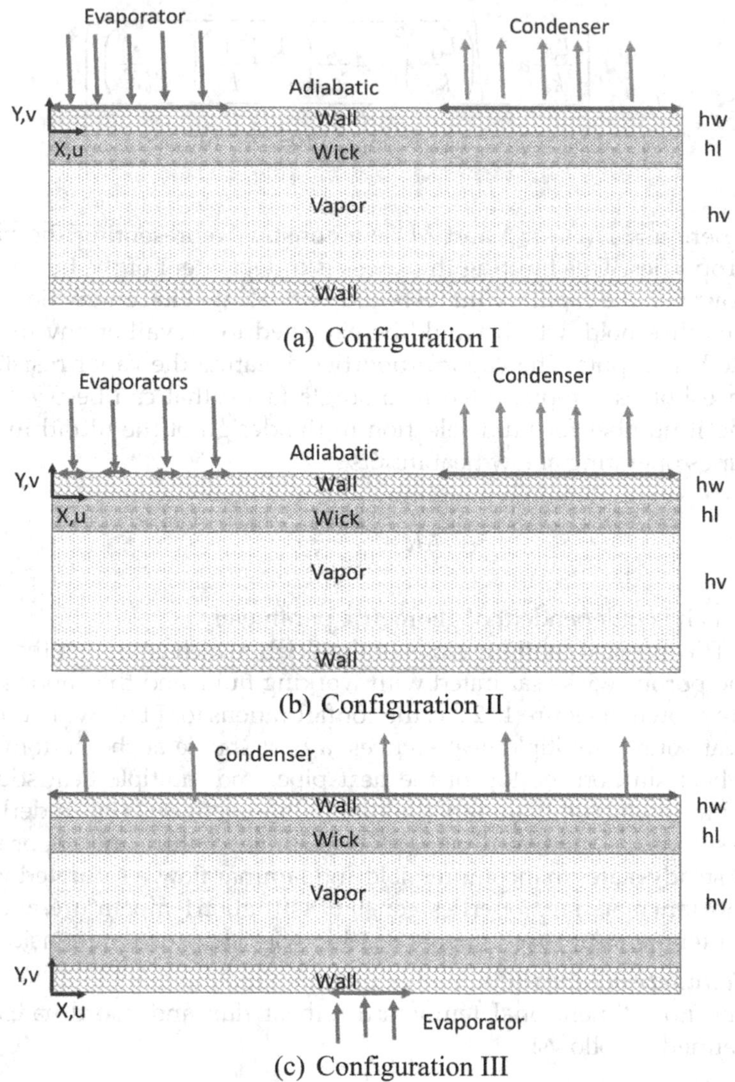

FIGURE 4.42
Flat heat pipe with different heating/cooling configurations: (a) configuration I, single heat source and sink at the top; (b) configuration II, multiple heat sources and sink at top; (c) configuration III, heat source at the bottom and heat sink on the top.

The boundary condition at the cap ($X = 0$ and $X = 1$) are as follows:

$$\frac{\partial \theta}{\partial X} = 0 \tag{4.100}$$

At the wick/wall interface ($Y = 0$), the thermal boundary condition can be determined by assuming linear temperature profile across the thin wick structure and constant saturation temperature at the liquid–vapor interface. The boundary condition can be written as follows:

$$\frac{\partial\theta}{\partial Y}=\frac{k_{\text{eff}}H_{\text{w}}}{k_{\text{w}}H_{\text{l}}}\theta \tag{4.101}$$

The boundary condition at the outer wall ($Y = 0$) are constant heat fluxes in the evaporators and the condenser, which can be located at different points on the heat pipe, and a heat flux equal to zero in the adiabatic sections at the rest of the surface. This is shown as follows:

$$\frac{\partial\theta}{\partial Y}=\begin{cases}1 & \text{evaporator}\\ 0 & \text{adiabatic}\\ -\gamma & \text{condenser}\end{cases} \tag{4.102}$$

The nondimensional temperature for configurations I and II can be expanded in the form of an infinite Fourier series as follows:

$$\theta(X,Y)=\sum_{m=1}^{\infty}A_m(Y)\cos(m\pi X) \tag{4.103}$$

The nondimensional input heat flux is dependent on the location of heat sources and sinks, and can be written as follows:

$$\frac{\partial\theta}{\partial Y}=\sum_{m=1}^{\infty}B_m\cos(m\pi X) \tag{4.104}$$

where

$$B_m=\frac{2}{m\pi}\{\sin(m\pi L_{2e})-\sin(m\pi L_{1e})+\sin(m\pi L_{4e})-\sin(m\pi L_{3e})+\sin(m\pi L_{6e}) \\ -\sin(m\pi L_{5e})+\sin(m\pi L_{8e})-\sin(m\pi L_{7e})-\gamma[\sin(m\pi L_{2c})-\sin(m\pi L_{1c})]\} \tag{4.105}$$

where $L_{1e}, L_{2e}, L_{3e}, L_{4e}, L_{5e}, L_{6e}, L_{7e}, L_{8e}, L_{1c}$ and L_{2c} express the locations of the heat sources and heat sink.

Substituting (4.103) and (4.104) into (4.99), (4.101) and (4.102) yields the following expression for coefficient A_m:

$$A_m(Y)=B_m\frac{\left(m\pi+\frac{k_{\text{eff}}}{k_{\text{w}}H_{\text{l}}}\right)\exp(m\pi H_{\text{w}}Y)+\left(m\pi-\frac{k_{\text{eff}}}{k_{\text{w}}H_{\text{l}}}\right)\exp(-m\pi H_{\text{w}}Y)}{m\pi H_{\text{w}}\left[\left(m\pi+\frac{k_{\text{eff}}}{k_{\text{w}}H_{\text{l}}}\right)\exp(m\pi H_{\text{w}}Y)-\left(m\pi-\frac{k_{\text{eff}}}{k_{\text{w}}H_{\text{l}}}\right)\exp(-m\pi H_{\text{w}}Y)\right]} \tag{4.106}$$

- ***Vapor flow***
  A parabolic velocity profile is used for vapor flow within the heat pipe. The velocity distribution is represented by a functional product in the x- and y-directions:

$$u_v(x,y) = -6U_v(x)\left[\left(\frac{h_l}{h_v}\right)^2 + \frac{h_l}{h_v} + \left(1+2\frac{h_l}{h_v}\right)\left(\frac{y}{h_v}\right) + \left(\frac{y}{h_v}\right)^2\right] \quad (4.107)$$

where $U_v(x)$ is the local mean velocity along the x-axis.

The continuity equation for the two-dimensional incompressible vapor flow can be integrated with respect to y to determine $U_v(x)$

$$\int_{-(h_l+h_v)}^{-h_l}\left(\frac{\partial u_v}{\partial x} + \frac{\partial v_v}{\partial y}\right)dy = \int_{-(h_l+h_v)}^{-h_l}\frac{\partial u_v}{\partial x}dy + v_{v,I}(-h_l) - v_v(-(h_l+h_v)) \quad (4.108)$$

where $v_{v,I}$ denotes the vapor interfacial velocity. The heat flux normal to the liquid–vapor interface, q_I can be calculated from the conduction model with the assumption of linear temperature profile across the wick:

$$v_{v,I}(-h_l) = \frac{q_I}{\rho_v h_{fg}} = q_e \frac{k_{eff}h_w}{\rho_v h_{fg} k_w h_l}\sum_{m=1}^{\infty} A_m(0)\cos\left(\frac{m\pi x}{l}\right) \quad (4.109)$$

$$v_v(-(h_l+h_v)) = 0 \quad (4.110)$$

Integrating equation 4.109 with respect to x and using the boundary condition at the beginning of the evaporator $u_v = 0$, will result in an expression for $U_v(x)$:

$$U_v(x) = q_e \frac{k_{eff}h_w}{\rho_v h_{fg} k_w h_l}\sum_{m=1}^{\infty}\left(\frac{l}{m\pi}\right)A_m(0)\sin\left(\frac{m\pi x}{l}\right) \quad (4.111)$$

Therefore, the quasi-two-dimensional velocity profile is as follows:

$$u_v(x,y) = -6q_e \frac{k_{eff}h_w}{\rho_v h_{fg} k_w h_l}\sum_{m=1}^{\infty}\left(\frac{l}{m\pi}\right)A_m(0)\sin\left(\frac{m\pi x}{l}\right) \times\left[\left(\frac{h_l}{h_v}\right)^2 + \frac{h_l}{h_v} + \left(1+2\frac{h_l}{h_v}\right)\left(\frac{y}{h_v}\right) + \left(\frac{y}{h_v}\right)^2\right] \quad (4.112)$$

The boundary layer form of the x-momentum equation for steady-state incompressible laminar vapor flow can be integrated to obtain the vapor pressure distribution:

$$\rho_v \int_0^x \int_{-(h_l+h_v)}^{-h_l} \frac{\partial (u_v)^2}{\partial x} dx\,dy + \rho_v \int_0^x \int_{-(h_l+h_v)}^{-h_l} \frac{\partial (u_v v_v)}{\partial y} dx\,dy$$
$$= \int_0^x \int_{-(h_l+h_v)}^{-h_l} -\frac{\partial p_v}{\partial x} dx\,dy + \mu_v \int_0^x \int_{-(h_l+h_v)}^{-h_l} \frac{\partial^2 (u_v)}{\partial y^2} dx\,dy \tag{4.113}$$

Assuming the constant vapor pressure at each cross-section, and utilizing the boundary conditions at the beginning of the evaporator, $u_v = 0$ and $p_v = p_{ref}$ yields

$$p_v = p_{ref} + q_e \frac{k_{eff} h_w}{\rho_v h_{fg} k_w h_l h_v^3} \sum_{m=1}^{\infty} A_m(0)\left[\cos\left(\frac{m\pi x}{l}\right) - 1\right] + 18\rho_v U_v^2 \left[\frac{1}{5}\left(\frac{h_l}{h_v}\right)^5 + \frac{1}{30}\right] \tag{4.114}$$

where U_v is obtained from 4.111.

The temperature drop in the vapor can be related to the pressure drop in the vapor region by applying the Clausius–Clapeyron equation and using the ideal gas law:

$$T_v = \frac{1}{\dfrac{1}{T_{ref}} - \dfrac{R}{h_{fg}} \log\left(\dfrac{p_v}{p_{ref}}\right)} \tag{4.115}$$

where R is the ideal gas constant, h_{fg} is the latent heat, and p_v is the pressure drop obtained from (4.112).

- ***Liquid flow***

The continuity equation for the incompressible liquid flow can be integrated with respect to y to obtain the liquid axial velocity $u_l(x)$:

$$\int_{-h_l}^{0} \left(\frac{\partial u_v}{\partial x} + \frac{\partial v_v}{\partial y}\right) dy = -v_{l,I}(-h_l) - h_l \frac{du_l}{dx} \tag{4.116}$$

where $v_{l,I}$ denotes the interfacial velocity (normal to the liquid–vapor interface) for the liquid and is related to the vapor interfacial velocity by

$$\rho_l v_{l,I} = \rho_v v_{v,I} \tag{4.117}$$

The axial liquid velocity can be calculated by integrating (with respect to x, using the velocity boundary condition at the beginning of the evaporator ($x = 0$), $u_l = 0$):

$$u_l(x) = q_e \frac{k_{eff} h_w}{\rho_v h_{fg} k_w h_l^2} \sum_{m=1}^{\infty} \left(\frac{l}{m\pi}\right) A_m(0) \sin\left(\frac{m\pi x}{l}\right) \tag{4.118}$$

The one-dimensional steady-state conservation of momentum for incompressible liquid flow in the wick can be expressed by Darcy's law, assuming negligible inertial effects in comparison to viscous losses, as follows:

$$\frac{dp_l}{dx} = -\frac{\mu_l u_l}{K} \tag{4.119}$$

where the permeability K is calculated for a mesh screen as follows:

$$K = \frac{d^2 \varphi^3}{122(1-\varphi)^2} \tag{4.120}$$

$$\varphi \approx 1 - \frac{1.05\pi N d}{4} \tag{4.118}$$

The liquid pressure is obtained by integrating and utilizing the pressure boundary condition at the end of the condenser ($x = l$), $p_l = p_v$:

$$\begin{aligned} p_l = {} & q_e \frac{\mu_l k_{eff} h_w}{\rho_l h_{fg} k_w K h_l^2} \sum_{m=1}^{\infty} \left(\frac{l}{m\pi}\right)^2 A_m(0) \left[\cos\left(\frac{m\pi x}{l}\right) - \cos(m\pi)\right] \\ & + p_{ref} + q_e \frac{12 \mu_v k_{eff} h_w}{\rho_v h_{fg} k_w h_v^3 h_l} \sum_{m=1}^{\infty} \left(\frac{l}{m\pi}\right)^2 A_m(0) \left[\cos(m\pi) - 1\right] \end{aligned} \tag{4.121}$$

- ***Capillary pressure***

$$p_{cap} = p_v - p_l \tag{4.122}$$

4.4.1.2 FEM Simulation

The schematic diagram of the simulated geometries is shown in Figure 4.42.

The properties of the working fluid (Table 4.4), copper and porous wick materials used in the vapor core chamber simulations are shown in the following table. Vapor properties are shown at temperature of 40°C.

The temperature fields predicted by the modeling approaches are shown in Figure 4.43; Figure 4.43a shows the temperature profile comparison along

TABLE 4.4

Properties of the Working Fluid, Copper, and Porous Wick Materials Used in the Heat Pipe Simulations

Property	Value
Copper density (ρ_{wall})	8,978 kg/m^3
Water liquid density (ρ_{water})	992.3 kg/m^3
Water vapor density (ρ_v)	0.05122 kg/m^3
Copper thermal conductivity (k_{wall})	400 W/mK
Wick effective thermal conductivity (k_{wick})	1.3 W/mK
Water vapor thermal conductivity (k_v)	0.05 W/mK
Copper specific heat capacity (C_p)	381 J/kgK
Water liquid specific heat capacity (C_p)	4,182 J/kgK
Water vapor specific heat capacity (C_p)	1,889 J/kgK
Water liquid viscosity (μ_l)	0.00065 Pa*s
Water vapor viscosity (μ_v)	0.0000096 Pa*s
Enthalpy of vaporization (h_{fg})	2,473 kJ/kg
Specific gas constant(*R*_const)	8.3145 J/molK
Water vapor mean molar mass (M_n)	0.018015 kg/mol
Wick porosity (φ)	0.5
Wick permeability (*K*)	$1 * 10^{-11}$ m^2

the *x*-direction of 50 W heat source between COMSOL and publication [107] and the error between them. Figure 4.43b shows the COMSOL simulated temperature profile with 30 W, 40 W, and 50 W heat source. The temperature profile has a maximum temperature at the evaporation side in the top surface and a minimum temperature at the condenser side. The plots reveal a high-accuracy match between COMSOL and theoretical solution. The maximum error is about 6% for configuration I and 0.075% for configuration II. From Figure 4.43b and d, we can see that by increasing 10 W at the heat source, the temperature gains about 3K for configuration I, and 4K for configuration II.

Figure 4.44 compares the velocities at the center of vapor chamber.

As we can see in the figure, the vapor has a maximum velocity at the vapor core center and reduce to 0 when it attaches the wall. For the COMSOL model, the maximum velocity is about 7 m/s, that's almost the same like the velocities of paper [107]. For configuration I, there is about 20% difference between analytical solution and COMSOL one, especially for the analytical solution, the velocity of configuration I has a non-symmetry form. But for configuration II, the difference in velocity is less than 4%. And by increasing 10 W of the heat source, the velocity gains about 1.5 m/s for configuration I and 6m/s for configuration II.

Finally, the most important part of the heat pipe is the capillary pressure in the vapor/wick interface. Figure 4.45 shows the capillary pressure at the wick/vapor interface. At the evaporation side, the water is evaporated and the vapor has a maximum pressure gain at this side, but at the condenser

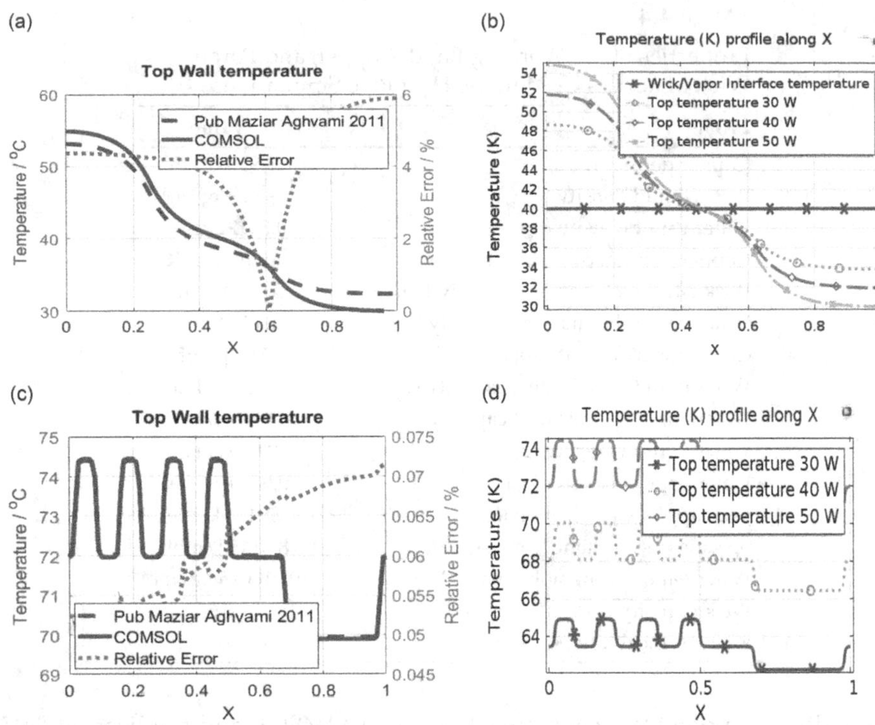

FIGURE 4.43
Temperature profile (a) comparison between COMSOL and pub with 50 W heat source [107] in MATLAB® configuration I, (b) parametrization of power in COMSOL configuration I, (c) comparison between COMSOL and pub with 50 W heat source [107] in MATLAB configuration II, and (d) parametrization of power in COMSOL configuration II.

side, the water is condensed and causes the water liquid's flow in the wick, so we can see a contrary process in the wick flow, the wick has a mass flow in at the condensation side, and has a mass flow out at the evaporation side, so the water flow from the condensation side to the evaporation side, but with a velocity much slower than the vapor core. From Figure 4.43, we can see that the capillary pressures of configuration I of COMSOL simulation and publication [107] have at maximum 20% difference. But for configuration II, the capillary pressures are almost the same. And the error is less than 0.5%.

A stationary model for vapor chamber operation was developed that allows for multiple, arbitrarily shaped, heat inputs on the evaporator-side face; the model predicts 2D fields of temperature, pressure, and velocity in the vapor chamber, subject to assumptions regarding the vapor chamber size, liquid and vapor flow, and the material properties. The governing mass, momentum, and energy equations in the wall, wick, and vapor core domains, were established. The model is compared with a numerical model [107].

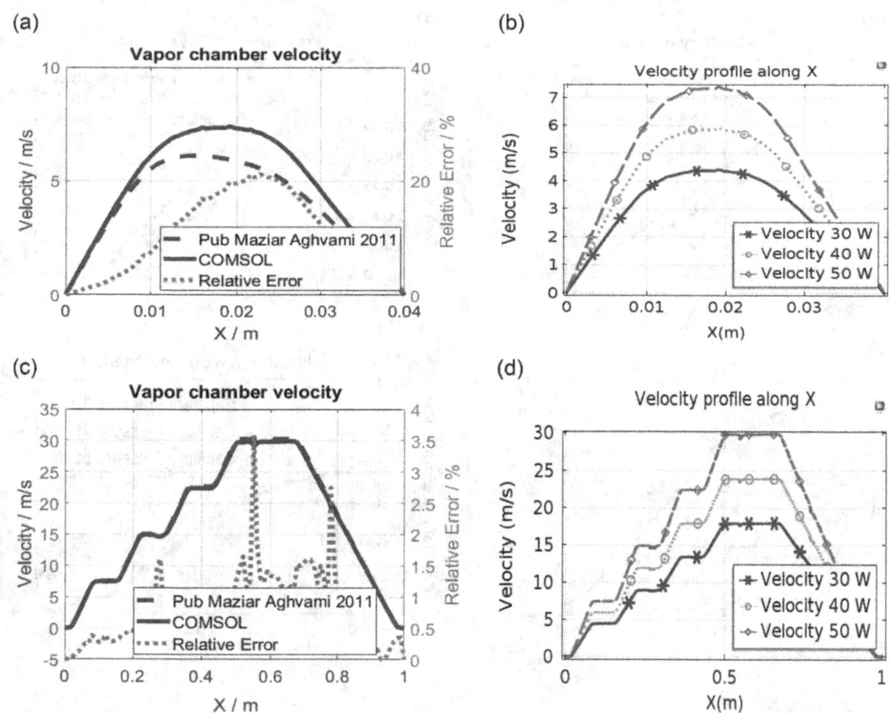

FIGURE 4.44
Velocities at the center of vapor chamber (a) comparison between COMSOL and pub with 50 W heat source in MATLAB configuration I, (b) parameterization of power in COMSOL configuration I, (c) comparison between COMSOL and pub with 50 W heat source [107] in MATLAB configuration II, and (d) parametrization of power in COMSOL configuration II.

4.4.2 Standalone Chip

The first step is to create the numerical model, representing a standalone chip [115], as shown in Figure 4.45. Those chips are not simply a stack of different layers but also include many small components (e.g., electrical interconnections) that would be unpractical to discretize. So, those components are homogenized with the material in their surroundings using the method proposed by de Crécy [116]. The planar and normal thermal conductivities are determined independently in order to represent the thermal behavior in those areas. Micro-heat pipe are located in the Si die up to the Si-molding junction. The model geometry, including the homogenized zones, is presented in Figure 4.46. Table 4.5 presents a summary of the thermal conductivities and dimensions used. The logic die thermal conductivity is approximated by a power law based on some empirical results (Figures 4.47 and 4.48).

The boundary conditions surrounding the chip are natural convection with a convection coefficient estimated to $h_{surr} = 10\ W/m^2K$ [115].

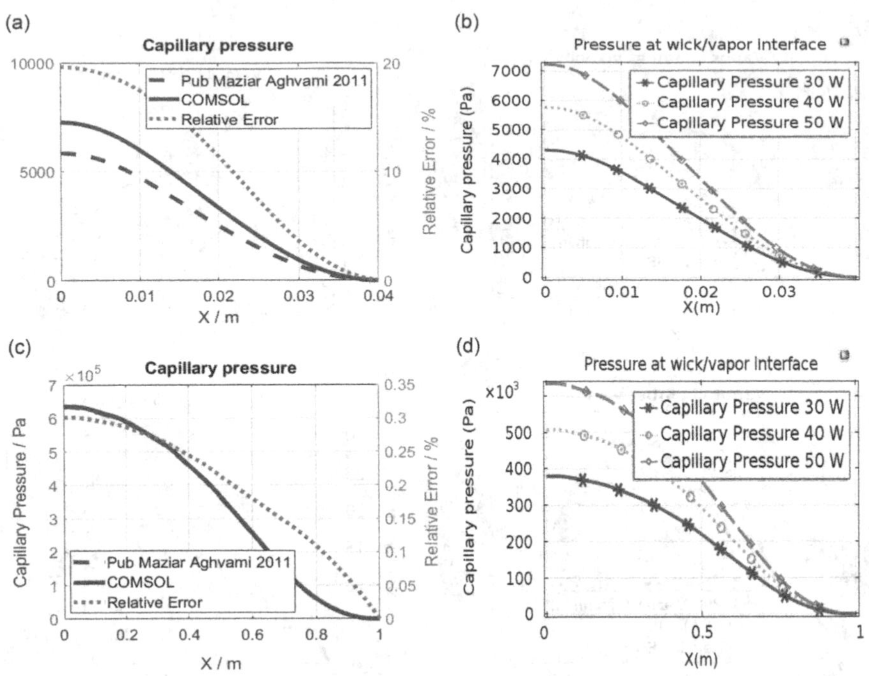

FIGURE 4.45
Capillary pressure at the wick/vapor interface (a) comparison between COMSOL and pub with 50 W heat source [107] in MATLAB configuration I, (b) parameterization of power in COMSOL configuration I, (c) comparison between COMSOL and pub with 50 W heat source configuration II, and (d) parameterization of power in COMSOL in MATLAB configuration II.

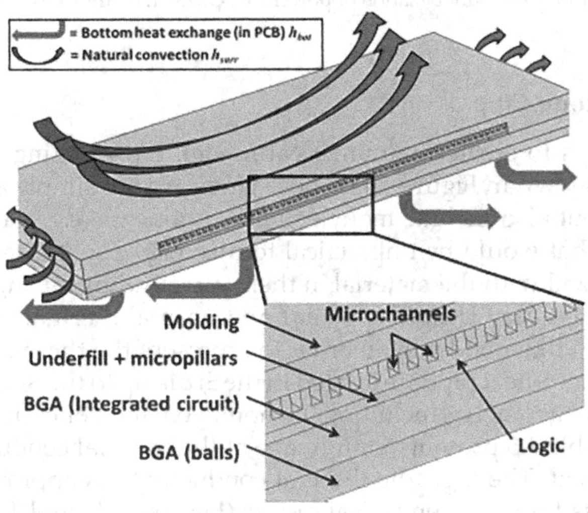

FIGURE 4.46
Simulated standalone chip.

TABLE 4.5

Chip Layer Dimensions and Thermal Conductivities

Layer	Thermal Conductivity (W/m-K)		Thickness (μm)	Sides (mm)
–	Planar	Normal	–	–
Molding	0.88		200 (above logic)	12 × 12
Si die	161,952 * $T^{-1.225}$		200	8.5 × 8.5
Underfill (polymer)	1.5		70	8.5 × 8.5
[a]Underfill (polymer + pillers)	1.9	3.5	70	
BGA (substrate)	92	0.6	288	12 × 12
BGA (balls array)	0.7	8	210	

T, Temperature (K).
[a] Homogenized zone.

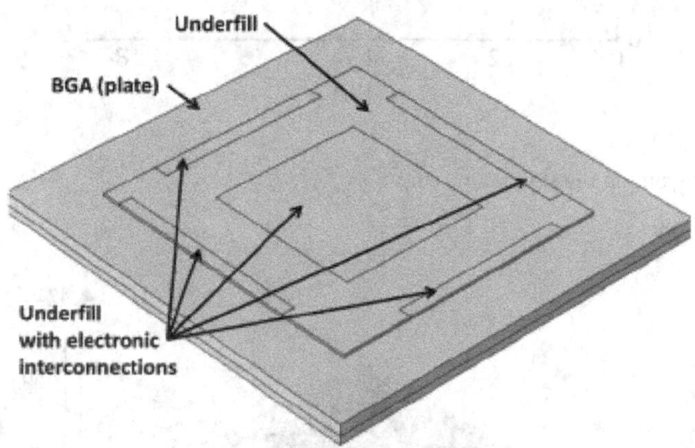

FIGURE 4.47
Simulated standalone chip showing the layer below logic die.

The bottom of the model is normally linked to a Printed Circuit Board (PCB), which allows important lateral heat spreading, with an equivalent convection coefficient estimated to h_{bot} = 570 W/m²K [115].

- ***Heat management***
  The heating power is distributed into eight cores of 360 × 490 μm² to represent the test-chip, as illustrated in Figure 4.49. With 2 W per core, a total of 16 W needs to be removed. In order to work properly, the chip maximum temperature must be below 125°C. Both the ambient and inlet temperatures are set at 40°C, which is an estimate for confined air.

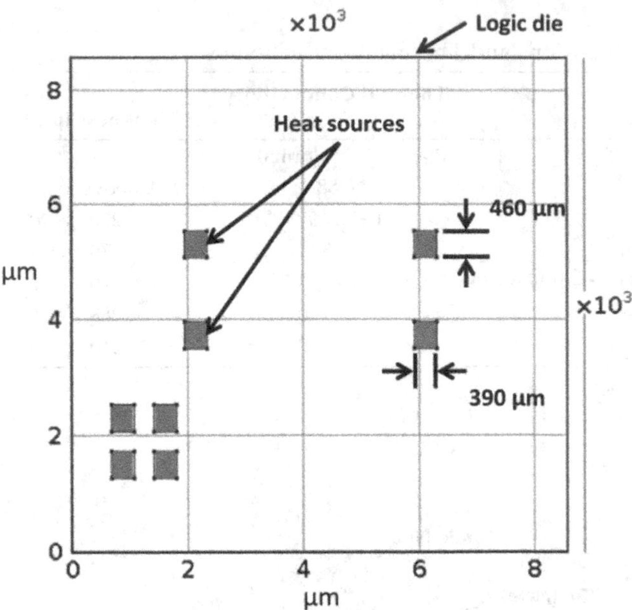

FIGURE 4.48
Heat sources layout on the standalone chip.

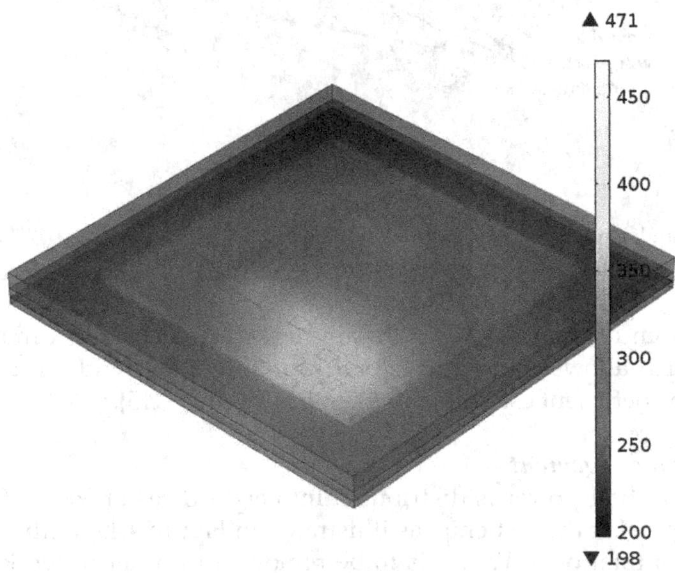

FIGURE 4.49
Chip simulation without micro-channels, extracting heat only by natural convection and PCB heat spreading.

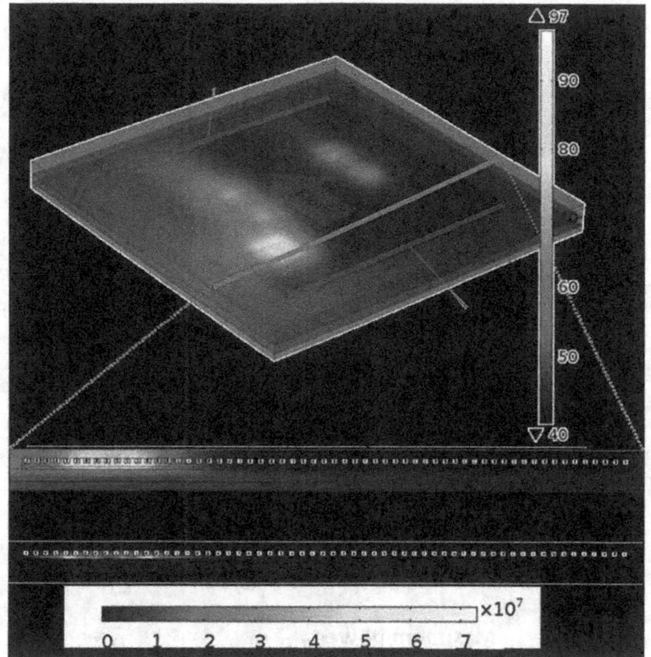

FIGURE 4.50
Micro-heat pipe simulation dissipating 16 W in a flip-chip with 85 μm long and 200 μm thick.

4.4.2.1 Results and Discussion

- ***Uncooled chip***
  A reference case is simulated without any micro-heat pipe. In this case, the chip is exclusively cooled by the spreading below the BGA and the natural surrounding convection. In Figure 4.49, the maximum temperature is well above the maximum allowed temperature of 125°C, which clearly shows the need of a cooling solution (Figures 4.50–4.53).
- ***Micro-heat pipe integration with standalone chip***
- ***Standalone chip application***

4.5 Conclusion

Efficient and perhaps elegant techniques to model electrothermal phenomena into subtracts, in 3D circuits, via frequency-dependent impedance extraction, are presented and programmed. They use a twofold approach:

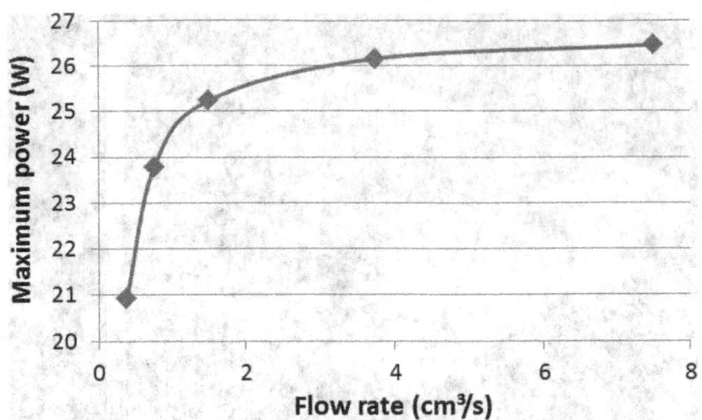

FIGURE 4.51
Maximum power admissible to reach a maximum temperature of 125°C according to heat pipe length-to-thick ratio.

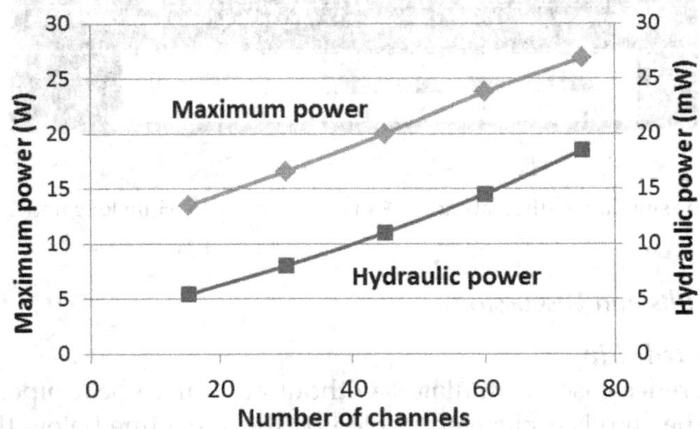

FIGURE 4.52
Maximum power admissible micro-heat pipe simulation with different porous thermal conductivity.

Green kernels and TLM method, as well as the use of FFTs. The speed of this latter technique makes it suitable for optimization of circuit layout, for minimization of substrate coupling related effects. However, the earlier models are often limited to the surface contacts.

In practice, in this work, contacts of any shape can be placed anywhere into the substrate; this is new, up to our knowledge.

We go on developing analytical models for some basic thermal analysis leading to a 3D IC temperature analytical model. Based on the correlated temperature model, study the influence of TSV's surface ratio r to the temperature, finally give the best r factor's range of eight dies 3D IC model. Based on the thermal resistance, we study the 3D CMP temperature characteristics

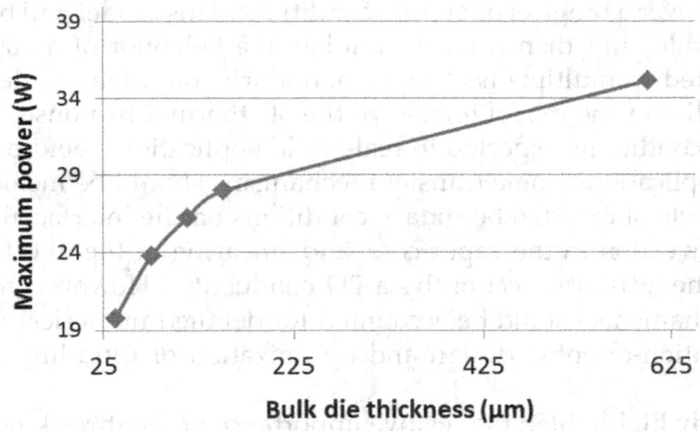

FIGURE 4.53
Maximum power admissible according to bulk die thickness.

and analyze the heat capacity, thermal resistance and heat dissipation's influence on chip temperature. As shown above, thermal resistance and power dissipation can affect 3D CMP's stationary temperature; so we can adopt different low power design to reduce chip's power dissipation, and we can also choose a higher heat conductivity material to reduce the thermal resistance to control the chip temperature. The higher heat capacity is, the longer time 3D CMP will need to get to the stationary temperature. However, the change of heat capacity will not change the final stationary temperature. Furthermore, this part gives a very important basis for the 3D CMP heat management.

In fact, this part tends to give a refined overview of 3D IC problems and a simple solution. We do not take so complicated case into consideration, but for further study, we will consider electrothermal effect of caches' and cores' interconnections.

Then, a substrate coupling simulation method suitable for execution in a conventional CAD environment is proposed. Moreover, to apply this approach to a realistic mixed signal design, several challenges have to be grasped. A key one concerns the way to obtain, for instance, the transient consumption current of an entire large digital part. Some CAD tools offer the possibility to simulate these currents.

Finally, introduced is some heat pipe modeling, necessary in recent up-to-date technologies (in our case: CMOS imagers), via an FEM-based model (cf. COMSOL Multiphysics), very accurate in the 2D and 3D simulations. Precise boundary conditions can be found; it considers the pressure field in the calculation, and especially it can extract the capillary pressure directly from the model, which will be very useful in the future study of wick layer's capillary effects. As future work, the simplification for the evaporated mass should be changed to a temperature-dependent parameters and a quantitative analysis of the model error, due to the simplifying assumptions employed, will be performed over a range of operating conditions. This model will be changed to a 3D model and then used to simulate the behavior of a vapor chamber subjected to multiple heat input boundary conditions to demonstrate the capability of the model to resolve the 3D thermal response to complex boundary conditions expected in real-world applications. Before coming to the real-applications, some transfer mechanisms should be included in the model: effects of external boundary conditions on the interfacial mass flux profile, convection in the vapor core, and smearing of the interfacial mass flux into the adiabatic section by a 2D conduction. However, these additional mechanisms should be accounted for detailed numerical models for the application-oriented design and optimization of ultrathin-form-factor heat pipes.

This study highlighted the relative importance of the three kinds of resistances met in microfluidic chip cooling, and that conventional thinking is challenged when realistic module configurations are considered.

5

Substrate Noise and Parasites: Toward 3D

5.1 Introduction

This last chapter is devoted to the backward noise and parasites. Noise is a highly prominent and stochastic phenomenon, but from the other point of view, parasitic effects can be determinist effects in the 3D IC design. This chapter gives more possibilities on the domain of the noise, rather subtle, which becomes prohibitive because of the miniaturization and the passage to the 2D through 3D, where reside strong conceptual problems. Noise is a universal phenomenon. Generally, micro-electronics and, of course, nano-electronics are domains where autonomous noise sources can be dependent on each other due to their proximity. Here, we will introduce concepts such as correlation of random variables, and noise spectral density. We will recall briefly the concept of transfer impedance; calculations at the one-dimensional level have already been developed, but there are few in 2D and nothing relevance in 3D. Velocity fluctuations or generation–recombination, which can help quantifying certain noises by using master equations like the Boltzmann equation, can be taken into account in statistical methods such as Monte Carlo.

Digital noise can affect analog parts of the circuit; for example, VCOs, whose phase noise can be crippling for digital modulations.

Laboratories work in the domain of background noise in semiconductors and devices, thus on fluctuation of physical or electrical quantities and *ad hoc* mathematical tools, thereby building a comprehensive and cooperative program via dialog between researchers, students, post-graduates, devices, and information. This covers fundamental studies as well as their application in modern submicronic devices (SiGe–C–) silicon transistors, ultimate CMOS, high-speed III–V devices (HEMT: High Electron Mobility Transistor), Heterojunction Bipolar Transistor, and optoelectronic devices (lasers, photodiodes).

In electronic chips, semiconductor devices are immersed in a more or less noisy environment. From a microscopic point of view, fluctuations of physical and electronic quantities can be apprehended, for instance, using the Boltzmann equation, numerically or via Monte Carlo methods, and by calculating, for example, the speed fluctuation correlation function and its Fourier transform, which will give the noise spectral density. It was Shockley [117], inventor of the bipolar transistor (1948), who initiated these studies. Van der Ziel [118,119] and Van Vliet [120], as well as other eminent researchers [121] contributed to these studies.

- *GR processes*

  The generation–recombination noise (G–R) represents a noise source that is typical and fundamental of a noise in semiconductor materials and related devices, where the carrier concentration can change the order of magnitude a lot. For electron density, fluctuations can happen between two levels: conduction band and donor impurity, or due to defects. Conductivity, associated with GR noise, can be written by relating it to an average carrier lifespan τ:

$$\sigma_{GR} = \frac{\sigma_0}{1 + j\omega\tau} \tag{5.1}$$

  A typical compatible BiCMOS 0.35 μm process is considered. A specific region (*P+/P*) is represented. The real profile is modeled by a stack of layers of uniform thickness and uniformly doped (see Figure 4.1).

  Let us consider two locations for this G–R—Figure 4.1a,b. We use, as for the electrical first order, or for thermal phenomena, the concept of spreading/transfer impedance (see Figure 5.2). Modulations of impedance resulting in these GR noises are represented in Figure 5.1. Our first results are qualitative, but also seem quantitatively realistic. For instance, we clearly see in both cases, the influence of the additional

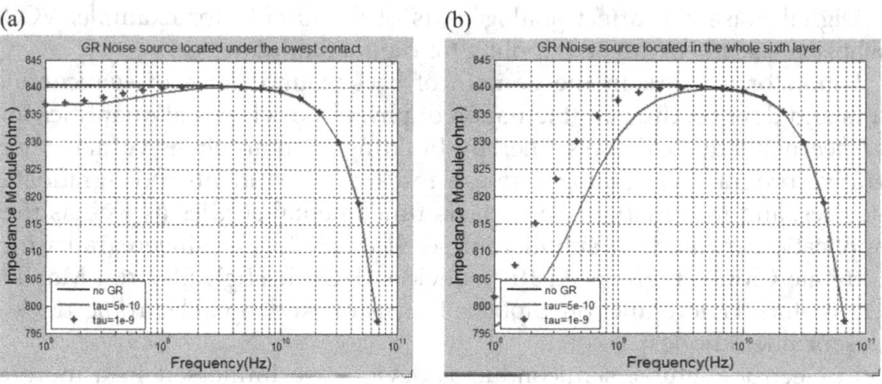

FIGURE 5.1
Introduction of a GR process: (a) defaults under contact C4 and (b) defaults in 6th layer (see Figure 4.1, Chapter 4).

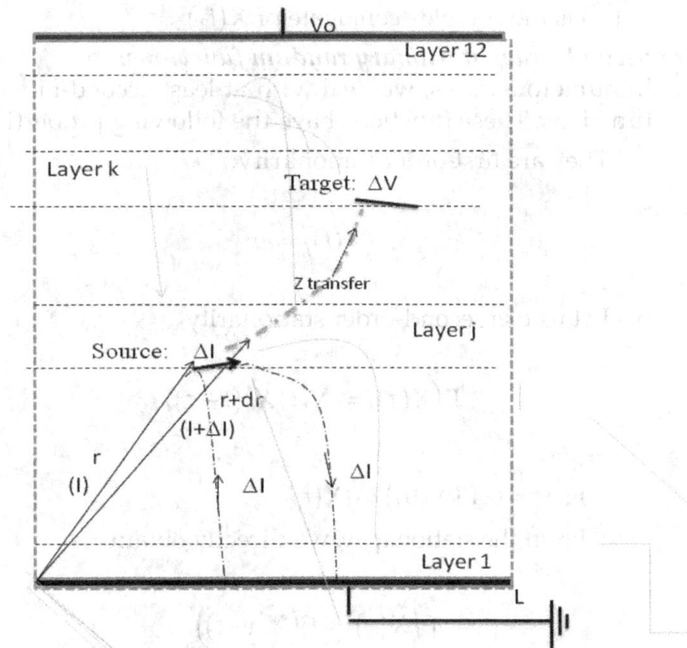

FIGURE 5.2
Schematic of the principle of the impedance field method $\left(Z = \dfrac{\Delta V_k}{\Delta I_j}\right)$.

GR current, reducing the resistance to low frequencies, before a cutoff frequency of $f_c = 1/(2\pi\tau)$. The impedance, for a life span of τ of 1 ns, can be reduced by 5% if its reference value, without GR.

- ***Random functions***
  A random variable is a real or complex number, the value of which depends on chance, (i.e., *a priori*, unpredictable) meaning that it depends on a "proof". A random function is one for which the value is constantly dependent on chance. A simple random function depends on one or more random variables. To know the value of the function at a specific moment *t*, then, there is a random variable *x*(*f*) that can be described by its probability density *p*(*x*, *t*) or, briefer still, by certain of its moments: $\langle X^1(t)\rangle$, $\langle X^2(t)\rangle$, etc. (average values are also called (life) expectations and marked *E*; thus: $E\{X(t)\}$, $E\{X^2(t)\}$ …)

  It will also be interesting to look at properties at two given instants: a moment that plays a specifically important role is covariance or correlation [122]:

$$\Gamma(X(t_1, t_2)) = \langle X(t_1)X^*(t_2)\rangle \tag{5.2}$$

$X^*(t_2)$ is the complex conjugate of $X(t_2)$

- ***Second-order stationary random functions:***
  In numerous cases, we deal with at least second-order random functions. These functions have the following properties [122]:
  a. They are first-order stationarity:

$$\langle X(t)\rangle = m \tag{5.3}$$

  b. Let us use second-order stationarity

$$\Gamma(X(\tau)) = \langle X(t)X^*(t-\tau)\rangle \tag{5.4}$$

  For $\tau = 0, \Gamma(X(0)) = \langle |X(t)|^2\rangle$
  From the stationarity, we directly obtain

$$\Gamma(X(\tau)) = \Gamma(X^*(-\tau)) \tag{5.5}$$

- ***Wiener–Khinchin theorem***
  If the signal can be written as a Fourier–Stieltjes transform, an integration of complex exponentials over the elementary increase $dx(\nu)$, can be expressed as follows:

$$X(t) = \int_{-\infty}^{\infty} e^{2i\pi\nu t}\, dx(\nu), \tag{5.6}$$

Taking (5.3) and (5.6) into consideration, the average, which is constant, can be written as follows:

$$m = \int_{-\infty}^{\infty} e^{2i\pi\nu t} \langle dx(\nu)\rangle \tag{5.7}$$

From which we obtain

$$\langle dx(\nu)\rangle = m\cdot\delta(\nu)d\nu \tag{5.8}$$

Thus, $\langle dx(\nu)\rangle$ is independent of ν and nonzero only when $\nu = 0$.

Now, if we take second-order stationarity into account, that is, if we substituted $X(t)$ and $X^*(t-\tau)$ with their values:

$$\Gamma_X(\tau)=\left\langle \int_{-\infty}^{+\infty} e^{2i\pi\nu t}\, dx(\nu)\cdot \int_{-\infty}^{+\infty} e^{-2i\pi(t-\tau)}\, dx^*(\nu') \right\rangle. \tag{5.9}$$

$$\Gamma_X(\tau)=\int_{-\infty}^{+\infty}\int_{-\infty}^{+\infty} \langle dx(\nu)dx^*(\nu')\rangle e^{2i\pi(\nu')\tau} e^{2i\pi(\nu-\nu')\tau} \tag{5.10}$$

$\Gamma_X(\tau)$ must be independent of *t*, and so all $\nu \neq \nu$ terms are zero. The integral (5.10) must be zero. Thus, we can write:

$$\langle dx(\nu)dx^*(\nu')\rangle = \delta(\nu-\nu')\gamma_X(\nu)d\nu d\nu' \tag{5.11}$$

Thus, there is no correlation between different frequencies for second-order stationary functions. The $\gamma_X(\nu)$ quantity is called spectral density of fluctuation of X. It is observed on the whole frequency spectrum, positive and negative. In general, we only take positive frequencies, and the measurable physical quantity $S_X(f)$ is defined by

$$S_X(f)=S_X(f)=\gamma_X(\nu=f)+\gamma_X(\nu=-f) \tag{5.12}$$

$$\Gamma_X(\tau)=\sum_{-\infty}^{+\infty}\sum_{-\infty}^{+\infty}\delta(\nu-\nu')\gamma_X(\nu)e^{2i\pi\nu'\tau}e^{2i\pi(\nu-\nu')\tau}\, d\nu\, d\nu' \tag{5.13}$$

After integrating over ν′, we obtain

$$\Gamma_X(\tau)=\int_{-\infty}^{+\infty} e^{2i\pi\nu\tau}\gamma_X(\nu)\, d\nu \tag{5.14}$$

$\Gamma_X(\tau)$ is the Fourier transform of the spectral density:

$$\gamma_X(\nu)=\int_{-\infty}^{+\infty} e^{-2i\pi\nu\tau}\Gamma_X(\tau)\, d\tau \tag{5.15}$$

Equations 5.14 and 5.15 relate to the Wiener–Khinchin theorem.

- ***Properties of spectral density of fluctuation.***
  a. Complex conjugate of spectral density

$$\gamma_X^*(\nu') = -\int_{-\infty}^{+\infty} \Gamma_X^*(-\theta)e^{-2i\pi\nu\theta}\, d\theta = \int_{-\infty}^{+\infty} \Gamma_x(\theta)e^{-2i\pi\nu\theta}\, d\theta = \gamma_X(\nu) \quad (5.16)$$

  Hence, $\gamma_X^*(\nu)$ is a real quantity. It is positive.

  b. White noise: the noise is said to be white if $\gamma_X(\nu)$ is constant, irrespective of ν: and, we have:

$$\Gamma_X(\tau) = \int_{-\infty}^{+\infty} Ae^{-2i\pi\nu t}\, d\tau = A\delta\tau \quad (5.17)$$

  When a noise is white, it means that successive events are not correlated. In practice, correlation is nonzero in a small time interval τ_0.

  τ_0 is the relaxation time of the system and we obtain: $\Gamma_X(\tau) = 0$ si $\tau > \tau_0$. In other words, $\gamma_X(\nu)$ is constant when $\nu < 1/\tau_0$.

5.2 Noise Calculation Methods

5.2.1 Langevin Method

- ***Principle of this Method [123]***

Of a very general scope, the Langevin method, which is applied to electronic noise, involves replacing the real device with a no-noise device in series with a noise source.

If η is the noise source, then the linear Langevin operator applied to the random variable is the said noise source, i.e.,

$$\hat{L}X(t) = \eta(t) \quad (5.18)$$

If the average value of $X(t)$ is zero, then we obtain

$$\hat{L}\langle X(t)\rangle = \langle \eta(t)\rangle = 0 \quad (5.19)$$

And

$$X(t) = \int_{-\infty}^{\infty} e^{2i\pi\nu t}\, dx(\nu) \quad (5.20)$$

$$\eta(t) = \int_{-\infty}^{\infty} e^{2i\pi\nu t}\, d\varsigma(\nu) \tag{5.21}$$

where $e^{2i\pi\nu t}$ is an eigenvector of $\hat{L}$, $Z(\nu)$ being its eigenvalue. Thus, we can write:

$$\hat{L}e^{2i\pi\nu t} = Z(\nu)\cdot e^{2i\pi\nu t} \tag{5.22}$$

Therefore, the noise spectral densities are connected via:

$$S_{\varsigma}(f) = \left|Z'(\nu, f)^2\right| S_x(f) \tag{5.23}$$

We see that the phase does not intervene here, with regard to second-order properties.

- ***Example: Generation–Recombination noise***
  In a semiconductor bar, we can generate ΔN number of new carriers, by lighting, for instance. If τ is the carrier lifetime, we can write:

$$\frac{d\Delta n}{dt} = -\frac{\Delta n}{\tau} \tag{5.24}$$

If we focus on fluctuations:

$$\frac{d\Delta n}{dt} = -\frac{\Delta n}{\tau} + N(t) \tag{5.25}$$

If the source is harmonic:

$$N(t) = N_0 e^{j\omega t} \tag{5.26}$$

Δn is also harmonic (cf. linearity):

$$\Delta n(t) = \Delta N e^{j\omega t} \tag{5.27}$$

Then, we obtain

$$S_N(f) = S_{\Delta n}(f)\left|\frac{1}{\tau} + j\omega t\right|^2 \tag{5.28}$$

$$S_{\Delta N}(f) = S_\eta(f)\frac{\tau^2}{1+\omega^2\tau^2}$$

If it is white noise,

$$\langle \Delta N^2 \rangle = \int_0^\infty S_{\Delta N}(f)df = S_\eta \tau^2 \int_0^\infty \frac{1}{1+4\pi^2 f^2\tau^2} df = S_\eta \frac{\tau}{4} \tag{5.29}$$

Thus:

$$S_{\Delta N}(f) = 4\langle \Delta N^2 \rangle \frac{\tau}{1+\omega^2\tau^2} \tag{5.30}$$

We know the spectrum if $\langle \Delta N^2 \rangle$ is white; we then use a master equation that regulates microscopic phenomena.

- ***Quantities associated to Noise in Electricity***
  Let us take the Langevin theorem again:

$$S_V(f) = |Z(f)|^2 S_I(f) \tag{5.31}$$

$$Z(f) = \frac{\Delta V}{\Delta I} \tag{5.32}$$

ΔV is complex voltage (cf. sinusoidal regime), superimposed at the quiescent point, continuous and induced by ΔI.

We can then define an equivalent noise temperature T_n, such as:

$$S_V(f) = 4kT_n(f)\mathrm{Re}(Z(f)) \tag{5.33}$$

but with

$$Z(f) = 1/Y(f) \tag{5.34}$$

Finally, we can define an equivalent noise current, I_n, in the following manner:

$$S_I(f) = 2qI_n(f) \tag{5.35}$$

Thus, spectral densities are linked to the square modulus of the complex impedance. This concerns macroscopic fluctuations. If the source term is known, the fluctuation of the quantity studied can be deduced. It is also noted that although the source has a white noise, a normal assumption, the variable studied does not, in general, have white noise; Z depends on the frequency.

- ***Observations:***
  1. To know one of the two terms, $I(f)$ or $S(f)$, we use microscopic conditions, via master equations (Boltzmann, Fokker–Planck equations, etc.),
  2. We get seen that it is absolutely necessary for the operator to be linear in order for the equation 5.18 to be applied; else, the Langevin description cannot be applied.

     It is only when L is linear that we get a noise source with an average value of zero, and thus find the no-noise system equation.

     If the system is not linear, equation 5.18 can be linearized, because, in general, we focus on the small fluctuations around the mean value. So, we can then use $X(t) = X_0 + \Delta X(t)$ and linearize equation 5.18, given that $\langle X_0 \rangle = 0$.
- ***Example application: Thermal noise of a Resistor:***
  Let us take a circuit (L, R); we find the spectral density of the noise current in the circuit, through which a zero average current flows. For this, we insert a source $\nu(t)$ in the perfect circuit and the circuit obeys the equation:

$$L\frac{di}{dt} + Ri = \nu(t) \tag{5.36}$$

We obtain

$$\nu = V_m e^{j\omega t} \tag{5.37}$$

$$i = I_m e^{j\omega t} \tag{5.38}$$

The result is well-known:

$$V_m = ZI_m \tag{5.39}$$

$$Z = R + j\omega L \tag{5.40}$$

Thus:

$$S_V(f) = |Z|^2 S_I(f) \tag{5.41}$$

$$S_I(f) = \frac{S_V(f)}{R^2 + L^2\omega^2} \tag{5.42}$$

Now, let us consider the microscopic aspect and calculate $\langle i^2 \rangle$. The average energy stored inside the coil is as follows:

$$\frac{1}{2}Li^2 \tag{5.43}$$

In 1D, we can write:

$$\frac{1}{2}Li^2 = \frac{1}{2}kT \tag{5.44}$$

and:

$$\langle i^2 \rangle = \int_0^\infty S_i(f)\, df = \int_0^\infty \frac{df}{R^2 + 4\pi^2 L^2 f^2} = \frac{S_V}{4RL} \tag{5.45}$$

So:

$$S_V = 4kRT\,(\text{Johnson–Nyquist theorem}) \tag{5.46}$$

- ***Forms of Langevin's equations***
  Langevin's equations often involve several coupled equations for $X_1(t)$, $X_2(t)$ variables, the average value of which satisfy the relationship of the form:

$$\frac{\partial \langle x_i(t) \rangle_{\text{cond}}}{\partial t} + \sum_i M_{ij} \langle x_i(t) \rangle_{\text{cond}} = 0 \tag{5.47}$$

$\langle x_i(t) \rangle_{\text{cond}}$ is a conditional mean respecting the given initial conditions. This is the case, for example, with electric circuits where initial conditions must be set; these equations are valid around the polarization point for all nonlinear systems. If derivatives of an order higher than 1 appear, the order of the system is to be reduced; for example, for a second-order system, we construct a new random variable, first derivative of the initial variable.

It can be shown that these phenomenological equations are equivalent to a Langevin description, which is the set of unconditional regression equations

The related Langevin description is as follows:

$$\frac{\partial \langle x_i(t) \rangle}{\partial t} + \sum_j M_{ij} \langle x_j(t) \rangle = \varsigma(t) \tag{5.48}$$

Given that the source terms are mean zero and have white noise.

- ***Example: Electron injection in p area of a PN junction.***
In an ohmic regime, the transport equation without generation–recombination: $-\partial n / \partial t$, is written as follows:

$$\frac{\partial\langle \Delta n(\vec{r},t)\rangle_{\text{cond}}}{\partial t} = -\frac{\langle \Delta n(\vec{r},t)\rangle_{\text{cond}}}{\tau} + \frac{1}{e}\nabla\langle \Delta\vec{j}_n(\vec{r},t)\rangle_{\text{cond}} \tag{5.49}$$

and also:

$$\langle \Delta\vec{j}_n(\vec{r},t)\rangle_{\text{cond}} = -e\mu_n\langle \Delta n(\vec{r},t)\rangle_{\text{cond}}\vec{E} + D_n\langle e\Delta n(\vec{r},t)\rangle_{\text{cond}} \tag{5.50}$$

$$\frac{\partial \Delta n(\vec{r},t)}{\partial t} = -\frac{\Delta n(\vec{r},t)}{\tau} + \frac{1}{e}\nabla\cdot\Delta\vec{j}_n(\vec{r},t) + \varsigma_n(\vec{r},t) \tag{5.51}$$

$$\Delta j_n(\vec{r},t) = -e\mu_n\Delta n(\vec{r},t)\vec{E} - eD_n\nabla\Delta n(\vec{r},t) + e\vec{\eta}_n(\vec{r},t) \tag{5.52}$$

Analogously, for holes:

$$\frac{\partial \Delta p(\vec{r},t)}{\partial t} = -\frac{\Delta p(\vec{r},t)}{\tau} - \frac{1}{e}\nabla\cdot\Delta\vec{j}_{np}(\vec{r},t) + \varsigma_p(\vec{r},t) \tag{5.53}$$

$$\Delta j_p(\vec{r},t) = e\mu_p\Delta p(\vec{r},t)\vec{E} - eD_p\nabla\cdot\Delta p(\vec{r},t) + e\vec{\eta}_p(\vec{r},t) \tag{5.54}$$

ς_n and ς_p are the noise sources.

5.2.2 The Impedance Field Method

- ***Principle of the method [119,124–130]***
The impedance field method was introduced by Shockley et al. [117]. This method differs from the Langevin method in the way it links the macroscopic quantity to elementary microscopic fluctuations. The spectral density of voltage fluctuations can be brought down to two basic calculations: that of the spectral density of voltage fluctuation and that of elementary fluctuations.
- ***Impedance fields:***
Let us consider a polarized specimen between electrode N and the grounding. If we inject a little current ΔI, flowing out of N, at a coordinate point $\vec{r}$ in the volume the potential of N varies by $\Delta V(N,\vec{r})$.

  Each noise course at the coordinate point $\vec{r}$ in the substrate (and also in the devices) produces a noise in another point of the volume and the surface (and electrodes).

The noise induced depends on its sources and its way of propagation via the impedance field. In order to study the noise between two contacts on the surface or in the substrate, we inject a little current (noise) that is complex, ΔI, at frequency f, at a coordinate point $r(\pm dr)$ in the substrate, superimposed on a continuous current between the contacts. This little current produces a potential $V(f)$ in another contact. Thus, we can write:

$$\Delta V(r+dr) = Z(r+dr)\delta I \tag{5.55}$$

$Z(r, f)$ is an impedance.

The resulting difference in voltage is as follows:

$$\delta V = \Delta V(r+dr) - \Delta V(r) \tag{5.56}$$

where Green's function ∇Z is called <u>the impedance field</u>.

$\delta I dr$ can be described by the current density $\delta j(r)$ in an infinitesimal volume dr^3:

$$\delta I dr = \delta j(r) dr^3 \tag{5.57}$$

δj is the current density (A m^{-2})

By injecting (5.55) in (5.56), we obtain:

$$\delta V_f(r) = \left(Z(r,f) - Z(r-dr,f)\right) \cdot \delta I = \nabla \cdot Z(r,f) dr \cdot \delta I \tag{5.58}$$

$$\delta V_f = \nabla Z(r,f) \cdot \delta \mathrm{j}(r) \cdot dr^3 \tag{5.59}$$

Equation 5.59 means that a current fluctuation in the dr^3 volume produces a voltage difference between two other contacts. Integrating on the domain considered, the total voltage produced by these currents is as follows:

$$\Delta V_f = \int \nabla Z(r,f) \cdot \delta \mathrm{j}(r) \cdot dr^3 \tag{5.60}$$

In order to explicitly calculate the noise density of $S_V(f)$, we project components, for example Cartesian, in x, y and z direction; in a direction, we obtain

$$\Delta V_f = \sum_{\alpha} \left[\int \frac{\partial Z(r,f)}{\partial \alpha} \nabla\, Z \cdot \delta j_{\alpha}(r) \cdot dr^3 \right] \tag{5.61}$$

Thus, the voltage noise spectral density at the junctions becomes

$$S_v(f) = 2\int_{-\infty}^{\infty} \Gamma_{\delta V}(N,\vec{r},\tau)\cdot e^{2\pi i f t}\,dt \tag{5.62}$$

where $\Gamma_{\delta V}(N,\vec{r},\tau)$ is the correlation function:

$$\Gamma_{\delta V} = \left(\delta V_f(t)\right)\cdot \delta V_f^*(t+\tau) \tag{5.63}$$

Density becomes:

$$S_v(f) = \sum_{\alpha}\sum_{\beta}\iint \frac{\partial Z'(\vec{r},f)}{\partial\alpha}\cdot\frac{\partial Z^*(\vec{r}',f)}{\partial\beta}.2\cdot\int_{-\infty}^{\infty}\overline{\delta j_\alpha(\vec{r},t)\delta j_\beta^*(\vec{r}',t+\tau)}\,e^{2\pi f i\tau}d\tau\cdot dr^3 dr'^3 \tag{5.64}$$

$$S_v(f) = \sum_{\alpha}\sum_{\beta}\iint \frac{\partial Z'(\vec{r},f)}{\partial\alpha}\cdot\frac{\partial Z^*(\vec{r}',f)}{\partial\beta}\cdot S_{j\alpha\beta}(\vec{r},\vec{r}',f)\,dr^3\,dr'^3 \tag{5.64bis}$$

$$S_v(f) = \iint \vec{\nabla} Z(\vec{r},f)\cdot Sj(\vec{r},\vec{r}',f)\cdot\vec{\nabla} Z^*(\vec{r}',f)\,dr^3\,dr'^3 \tag{5.65}$$

If K is a noise source:

$$S_{j\alpha\beta}(r,r'',f) = K_{\alpha\beta}(r,f)\delta(r'-r) \tag{5.66}$$

Equation 5.65 simplifies into:

$$S_v(f) = \int \nabla Z(r,f)\cdot K(r,f)\cdot\nabla Z^*(r'',f)\,dr^3 \tag{5.67}$$

- ***Calculation of impedance field***
  Calculating the impedance field is easy in the case of homogenous samples. But in the case of nonhomogeneous samples, in space charge regime, for instance, some precautions must be taken.
  a. We determine the working regime that is the polarization point, by
     - Solving local equations (Poisson's equation, electric neutrality, phenomenological expression of current density, etc.)
     - Taking boundary conditions into account, that is, for electrodes, to determine integration constants in the preceding differential equations.

Thus, we get the local mean field $E_0(r)$ based in the average current density j_0.

The average voltage at junctions in the sample is as follows:

$$V_0 = -\int_{N_0}^{N} \vec{E}_0(\vec{r})\cdot d\vec{r} \quad (5.68)$$

where N_0 and N are the electrodes. The average current is as follows:

$$I_0 = \iint \vec{j_0}\cdot\overrightarrow{D_S} \quad (5.69)$$

where S is a slice of the sample, for instance, the surface of an electrode.

b. We calculate the fluctuations of the electric field resulting from introducing current ΔI at coordinate point $\vec{r}$. This $\Delta\vec{E}(\vec{r},\vec{r}')$ field is obtained as follows:

$$\mathbf{E}(r') = \mathbf{E}_0(r') + \Delta\mathbf{E}(r,r') \text{ and } \mathbf{j}(r') = \mathbf{j}_0(r') + \Delta\mathbf{j}(r,r') \quad (5.70)$$

The Δ being considered as disturbances with regard to E_0 and I_0, these zero order terms give the equations for mean values, while the first-order terms in ΔE and ΔI give linear equations. Local Langevin equations thus obtained are solved and the integration constants are determined by imposing the same boundary conditions on $\Delta\vec{E}$ as those applied on $\mathbf{E}$. For $\Delta\mathbf{E}$, these conditions are at the level of electrode N and the $\bar{r}$ point, and at N and N_0 for $\mathbf{E}_0$. Thus, we get $\Delta\mathbf{E}(\mathbf{r}, \mathbf{r}')$

c. We then deduce the impedance gradient

- ***Application example: noise in ohmic space charge regime (simple injection)***
The sample is assumed to be a parallelepiped; we are dealing with a longitudinal unidimensional case. The space charge is not negligible and can make the device's characteristics nonlinear.
We distinguish between the two different regimes:
a. when the space charge of the injected carriers is very weak to significantly affect the volume's conductivity, we get the ohmic regime, where the mean current is proportional to the voltage at the junctions of the sample: $I_0 \propto V_0$.

$$j = -q\mu nE + \varepsilon\frac{\partial E}{\partial t} \quad (5.71)$$

$$\frac{\partial E}{\partial x} = \frac{-q}{\varepsilon}(\ n - N_D\ ) \tag{5.72}$$

$$\mathbf{j}_n(\vec{r},t) = -e\mu_n\left(\mathbf{E} + \frac{\varepsilon}{e}\frac{\partial \mathbf{E}}{\partial t}\right) + \varepsilon\frac{\partial \mathbf{E}}{\partial t} \tag{5.73}$$

$$(q = -e, > 0)$$

Consider harmonic fluctuations:

$$E = E_0 + \Delta E e^{jwt} \tag{5.74}$$

$$n = n_0 + \Delta n e^{jwt} \tag{5.75}$$

$$\Delta\mathbf{j}_n(\vec{r},t) = -eN_D\ \mu_n\left(\overline{\Delta E}\right) + \varepsilon\mu E_0\frac{\partial \Delta E}{\partial x} + \varepsilon\mu\frac{\partial \Delta E}{\partial t} + i\varepsilon\omega\mu\Delta\mathbf{E} \tag{5.76}$$

Hence:

$$\Delta j_n(\vec{r},t) = -eN_D\ \mu_n\Delta E + \varepsilon\mu E_0\frac{\partial \Delta E}{\partial x} + \varepsilon\mu\Delta E\frac{\partial E_0}{\partial x} + i\varepsilon\omega\mu\Delta E \tag{5.77}$$

If $n = N_D$ (e.g., at the ambient):

$$\varepsilon\mu E_0\frac{\partial \Delta E}{\partial x} + \left(-e\mu_n N_D + \varepsilon\mu\frac{\partial E_0}{\partial x} + i\varepsilon\mu\omega\right)\Delta E = \Delta j_n \tag{5.78}$$

$$\varepsilon\mu E_0\frac{\partial \Delta E(x,x')}{\partial x} + ((\ e\mu N_D + i\varepsilon\omega)\Delta E'x,x') = \Delta j_n \tag{5.79}$$

The solution is:

$$\Delta E(x,x') = \frac{\Delta j_n}{-e\mu_n + i\varepsilon\omega} + K(x)e^{-\frac{e\mu_n N_D + i\varepsilon\omega}{\epsilon\mu_0 E_0}x'} \tag{5.80}$$

$$\Delta E(x,x') = \frac{\Delta j_n}{-\ e\mu_n N_D + i\varepsilon\mu\omega} + \left(1 - e^{\frac{(-e\mu_n N_D + i\varepsilon\mu\omega)(x - x')}{\epsilon\mu_0 E_0}}\right) \tag{5.81}$$

The impedance gradient is defined by

$$\frac{\partial Z(N,x)}{\partial x} = -\frac{1}{\Delta I}\int_x^L \frac{\partial \Delta E(x,x'')}{\partial x}dx' \tag{5.82}$$

Then if the diffusion length is much smaller than that of the sample, we can write:

$$\frac{\partial Z(N,x)}{\partial x} = \frac{\Delta j_n}{S(\, i\varepsilon\mu\omega - e\mu_n)} \cdot \left(1 - e^{\frac{(-e\mu_n + i\varepsilon\mu\omega)(x-L)}{\epsilon\mu_0 E_0}} \right) \tag{5.83}$$

As for the noise,

$$\frac{\partial Z(N,x)}{\partial x} = \frac{\Delta j_n}{S(\, i\varepsilon\mu\omega - e\mu_n)} \tag{5.84}$$

Using the Einstein relationship,

$$\frac{kT}{q} = D/\mu \tag{5.85}$$

We obtain

$$S_v = 4e^2 D\mu \cdot \frac{nL}{(S\varepsilon\omega)^2 + (Se\mu N_D)^2} \tag{5.86}$$

Finally,

$$S_v = 4kT \frac{R}{1 + (\omega RC)^2} \tag{5.87}$$

It is the noise of the parallel *RC* circuit, where

$$C = \varepsilon S/L \tag{5.88}$$

$$R = L/(q\mu nS) \tag{5.89}$$

We can see in Figure 5.3 an *RC* model corresponding, for instance, to a multilayers substrate (see Chapter 4).

$$Z_{\text{eq}} = (1/R + jC\omega)^{-1} = R/(1 + jRC\omega)^{-1} \tag{5.90}$$

Thus:

$$S_v = 4kRT \frac{1}{1 + (\omega RC)^2} \tag{5.91}$$

We can also add a *RL* dipole (Figure 5.4: corresponding possibly to some bonding in the RF range).

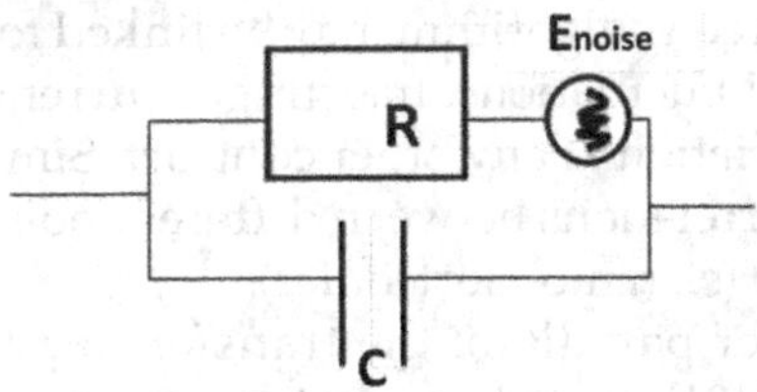

FIGURE 5.3
RC noise study circuit.

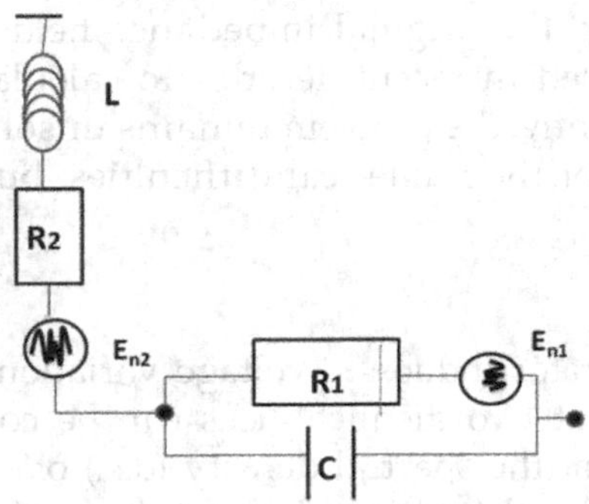

FIGURE 5.4
(*L*, *R*) and (*R*, *C*) noisy dipoles.

Then, we can write:

$$E_nE_n^* = E_{n2}E_{n2}^* + E_{n1}E_{n1}^* \cdot \frac{1}{1+(\omega R_1C)^2} + \left(E_{n2}E_{n1}^* \cdot \frac{1}{1-j\omega R_1C}\right) + E_{n1}E_{n2}^* \cdot \frac{1}{1+j\omega R_1C} \tag{5.92}$$

If the two noise processes are independent (hypothesis to be discussed for the nanoscale); and thermal:

$$E_{n1}E_{n2}^* = E_{n2}E_{n1}^* = 0 \tag{5.93}$$

Then, the total noise is:

$$S_{E_n} = 4kR_1T\frac{1}{1+(\omega R_1C)^2} + S_{E_{n2}E_{n2}} = 4kR_1T\frac{1}{1+(\omega R_1C)^2} + 4kR_2T \tag{5.94}$$

5.2.3 Transfer Impedance Method

We saw, in the impedance field method, that a current generator in a *dx* slice can induce a voltage variation anywhere in the whole device [119,125–130].

As we suggested, Green's algorithm may be linked to a transfer impedance, considering the signal fluctuations. Injecting a current variation to a contact creates a potential variation at any other contract. Similarly, we must be able to analyze possible correlations between different noise sources autonomous *ab initio* but very close (cf. nano-electronics).

We consider another paradigm: the transfer impedance method (TIM); between two points [131], two slices, or two volume elements, via a linear relationship between the electric field response or the potential difference at a current disturbance, situated far away. Hence, we get the possibility of calculating internal noise spectra that comes from these fluctuations. This method, generalizing the original impedance field method of Shockley et al., was widely used in recent decades to calculate noise and is only efficient in 1D. Currently, the problem remains unsolved in 3D (even often in 2D), not because of the numerical difficulties, but in a physical point of view.

- ***Principle***

  A current generator induces a voltage variation in every dx slice. The result is that two distinct slices can be correlated. Thus, we cannot write that the spectral density (d.s.) of voltage at junctions of the whole set is the sum of d.s. of voltage of each separate slice (cf. Salami method).

  The $\Delta v(x), \Delta v(x')$ cross terms must be added. More specifically, let us consider two slices, numbered as α and β. Microscopic noise sources, each sheltered by a noise current generator marked Si_α and Si_β, are untiled, but not the voltage fluctuations induced by these noise sources. Voltage induced at junctions of slice β by current i_α is $Z_{\beta\alpha}(w) i_\alpha$ where $Z_{\beta\alpha}$ is the transfer impedance between slices α and β. As a consequence, the total voltage induced at the junctions of β for all the current generators of different slices is as follows:

$$\Delta V_\beta = \sum_\alpha z_{\beta\alpha} i_\alpha \tag{5.95}$$

i_α and i_β are uncorrelated; we obtain

$$S_v \Delta f = \sum_{\alpha\gamma}^{\infty} \left|\overline{i_\alpha}^2\right| z_{\beta\alpha} z_{\alpha\gamma}^* \tag{5.96}$$

$$\vec{E}(\vec{r}) = \vec{E_0}(\vec{r}) + \delta\vec{E}(\vec{r}, \omega) e^{i\omega t} \tag{5.97}$$

$$\vec{j}(\vec{r}) = \vec{j_0}(\vec{r}) + \delta\vec{j}(\vec{r}, \omega) e^{i\omega t} \tag{5.98}$$

The first-order terms are linked via an operator $\hat{L}$:

$$\overleftrightarrow{L}\,\delta\vec{E}(\vec{r},\omega)=\delta\vec{j}(\vec{r},\omega) \tag{5.99}$$

We can then introduce Green's functions of the operator $\hat{L}$

$$\overleftrightarrow{L}\,\vec{\vec{Z}}\left(\vec{r},\vec{r'},\omega\right)=\vec{\vec{I}}\cdot\delta\left(\vec{r}-\vec{r'}\right) \tag{5.100}$$

where $\vec{\vec{I}}$: unit matrix

$$\delta\vec{E}(\vec{r},\omega)=\iiint\vec{\vec{z}}\left(\vec{r},\vec{r'},\omega\right)\delta\vec{j}(\vec{r},\omega)d^3r' \tag{5.101}$$

The ddp at junctions of a "*dr*" on an inter-electrode line can be written as follows:

$$\delta^2V_{dr}=-dr^T\iiint\vec{\vec{z}}\left(\vec{r},\vec{r'},\omega\right)\delta\vec{j}(\vec{r},\omega)d^3r' \tag{5.102}$$

And so, between two electrodes:

$$\delta V=-\int_O^L dr^T\iiint\vec{\vec{z}}\left(\vec{r},\vec{r'},\omega\right)\delta\vec{j}(\vec{r},\omega)d^3r' \tag{5.103}$$

Finally, for a material of L length, after some simple calculation that may be a little "heavy", we obtain

$$S_v\Delta f(\omega)=\int_0^L\int_0^L\iiint d^3r''\iiint\vec{\vec{z}}\left(\vec{r},\vec{r''},\omega\right)\vec{j_N}\left(\vec{r},\vec{r''},\omega\right)d^3r'''$$

$$\cdot dr^T\cdot\vec{j_N}\left(\vec{r},\vec{r'''},\omega\right)d^3r'\,\vec{\vec{z}}\left(\vec{r'},\vec{r'''},\omega\right) \tag{5.104}$$

We also have the spectral density of current fluctuation:

$$S_i\left(\vec{r'},\vec{r''},\omega\right)=\overline{J_N}\left(\vec{r'},\omega\right)\cdot\overline{J_N}^T\left(\vec{r''},\omega\right)=K\left(\vec{r'},\omega\right)\cdot\delta\left(\vec{r''},\vec{r'}\right)\Delta f \tag{5.105}$$

K is the noise source.

If we want to get voltage drops or impedance of the whole sample; curvilinear integrals must be calculated. Currently, some calculations can be carried out only in 1D, and this is a huge hurdle.

Equation 5.104 is written as follows:

$$S_v(\omega) = \iint \vec{\nabla} Z(r,\omega) \cdot Sj(r,\omega) \nabla Z^T \, dr^3 \tag{5.106}$$

(to be compared with the impedance field method).

Note: $S_v(\omega)$ is zero if the Z gradient is zero. Hence, we can create a (straight) strongly and uniformly doped slice between the noise source and a target to be protected.

- ***One-dimensional structures: Noise of a bar in ohmic regime in high frequencies.***

$$E = E_0 + \Delta E e^{jwt} \tag{5.107}$$

$$n = n_0 + \Delta n e^{jwt} \tag{5.108}$$

$$j = j_0 + \Delta J e^{jwt} \tag{5.109}$$

$$j = -q\mu n E + \varepsilon \frac{\partial E}{\partial t} \tag{5.110}$$

$$\frac{\partial E}{\partial x} = \frac{-q}{\varepsilon}(n - N_d) \tag{5.111}$$

In high frequencies, we can write:

$$\frac{\delta v}{\delta E} = \frac{\mu_0}{1 + i\omega\tau} \tag{5.112}$$

Note: In very low frequencies, $\frac{\delta v}{\delta E}$ tends toward μ_0 and in ohmic regime $\frac{\partial E}{\partial x} = 0$, thus, $n = N_d$

$$\delta I = -qA(n_0 \delta v + v_0 \delta n) + ieA\omega\delta E \tag{5.113}$$

where A is a square section.

$$\delta I = \left(\frac{qAn_0\mu_0}{1 + i\omega\tau} + i\omega A\right)\delta E - qAv_0\delta n \tag{5.114}$$

$$-\epsilon A\mu_0 E_0 \frac{\partial z(xx',\omega)}{\partial x} + \left(\frac{qAn_0\mu_0}{1 + i\omega\tau} + i\omega\epsilon A\right) z(x,x',\omega) = \delta'(x - x') \tag{5.115}$$

As for transfer impedance:

$$\in A\mu_0 E\frac{\partial z(x,x,\omega)}{\partial x}+\left(\frac{qAn_0\mu_0}{1+i\omega\tau}+i\omega\in A\right)z(x,x,\omega)=\delta'(x-x') \quad (5.116)$$

The solution for this differential equation is then:

$$z(x,x',\omega)=Ke^{\frac{\left(qn_0\mu_0-\omega^2\tau\in+\ i\omega\in\right).x}{\in\mu_0 E_0}} \quad (5.117)$$

Via, for example, the variation of constants theorem, we get K

$$K=-\frac{1+i\omega\tau}{AE_0\in\mu_0}+i\omega\in Ae^{-\lambda x'}H(x-x')+C \quad (5.118)$$

where H is the Heaviside step function;

And:

$$\lambda=\frac{qn_0\mu_0-\omega^2\tau\in+i\omega\in}{\in_0\mu_0 E_0} \quad (5.119)$$

Introducing δI in x' results in $x > x'$ if we ignore the diffusion.

In short:

$$z(x,x',\omega)=-\frac{1+i\omega\tau}{A\in_0\mu_0 E_0}e^{\frac{\left(qn_0\mu_0-\omega^2\tau\in+i\omega\in\right)(x-x')}{\in\mu_0 E_0}}H(x-x') \quad (5.120)$$

Hence, we have its gradient:

$$\nabla z(x',\omega)=\frac{1+i\omega\tau}{A\in_0\mu_0 E_0}\left(1-e^{\frac{\left(qn_0\mu_0-\omega^2\tau\in+\ i\omega\in\right)(L-x')}{\in\mu_0 E_0}}\right) \quad (5.121)$$

In ohmic regime (E constant, very weak): $\lambda\to\infty$ we obtain

$$\lim_{E_0\to 0}\vec{\nabla}z(x',\omega)=\frac{1+i\omega\tau}{A\left(qn_0\mu_0-\omega^2\tau\in+i\omega\in\right)} \quad (5.122)$$

Diffusion noise power is thus:

$$S_V=4q^2A\int_0^L n_0 D_0\left|\nabla Z\right|^2 dx \quad (5.123)$$

Yet, in ohmic regime: $kT/q = D_0/\mu_0$

$$S_V = 4\frac{kT}{Aqn_0\mu_0}L = 4kT\frac{L}{\sigma_0 A} = 4kRT \tag{5.124}$$

σ_0 is the ohmic conductivity

We get the well-known Nyquist–Johnson relationship.

Note: The approach for calculating the impedance field, in practice, can be to force the AC sources at each mesh node of the device and evaluate the response at the device's electrodes.

A method derived from the system principle is simpler to implement; we calculate the trans-impedance Z_{ij}, where i is any node and j is an electrode. As a consequence, $Z_{ij} = Z_{ji}$; it can be checked by applying an AC source at a terminal and calculating the internal potentials at each node.

- ***GR defaults***

  Finally, Figure 5.5 presents the noise spectral densities of previous two GR cases. For this, we use the impedance transfer paradigm, applied to the multilayer substrate (cf. Figure 4.1b), with GR defaults in the 6th layer, using equation 5.1.

  It should be possible to study current noise, induced by temperature fluctuations in the volume, with these methods.

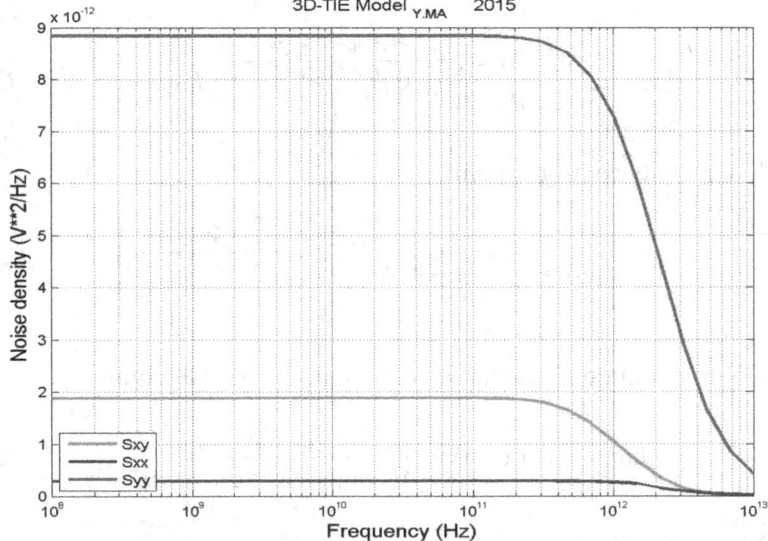

FIGURE 5.5
Noise spectral densities.

5.3 Digital Perturbations

5.3.1 Introduction

More and more system-on-chip design requires the integration of both very fast analog circuits and large digital blocks. Up to now, the perturbations introduced by the digital part against the analog one are a key problem, which is hard to handle by the designers, especially in the radio-frequency (RF) range. Substrate noise involved in by the digital part will degrade the performance of analog circuits. The in-depth problem is to construct efficient models for the digital part activity, the substrate, and their coupling, in a package that could be embedded in some unique computer aided design (CAD) tool. The well-known tool SPICE is able to model such phenomena: substrate or digital signals; but for a complex technology, with very large digital blocs, it seems too optimistic. Some significant attempts on modeling noise generation injected in the bulk, on a "SPICE" and a process/device point of view have been proposed; but it seems accurate for relatively small circuits and simple parasitic waveforms, traveling through the substrate.

The idea described below has been first proposed in [133–138]; we work out some developments, and finally present, for the moment, simple examples for the practical application of the methodology we did implemented.

5.3.2 Methodology

Our subjects of interest are the digital transmission/propagation systems.

For a general approach, consider M specific noise (parasitic) sources, $x_1(t)$, $x_2(t)$, …, $x_M(t)$, time-continuous signals, modulated by an encoded message $c[n]$. It can be written in the general form:

$$x(t)=\sum_{n=-\infty}^{\infty} g^{c[n]}(t-nT) \tag{5.125}$$

This equation means that the "symbol" $c[n]$, the switching event, emitted at the instant nT, assigns the waveform $g^{c[n]}(t-nT)$, where T is the period of the digital clock. Let s_n be a stationary Markov chain, with state set $S=\{\sigma_1,\sigma_2,\ldots,\sigma_I\}$ and $c[n]=f(s_n)$ be a vector-valued stochastic process obtained as a memoryless function of s_n. We assume that the Markov chain s_n is ergodic (s_n has only one class of communicating states. No degeneracy is involved in this process; a temporal average is equal to an ensemble one).

$g(t)$ is a $M \times I$ matrix, which takes the general form:

$$\mathbf{g}(t) = \begin{bmatrix} g_1(t,\sigma_1) & g_1(t,\sigma_2) & . & . & g_1(t,\sigma_I) \\ . & . & . & . & . \\ . & . & . & . & . \\ . & . & . & . & . \\ g_M(t,\sigma_1) & g_1(t,\sigma_2) & . & . & g_M(t,\sigma_I) \end{bmatrix} \tag{5.126}$$

Then $c[n]$ can be written $c[n] = S_nC$, where C is an $M \times I$ matrix, and where $S_n = [S_n(1), S_n(2), \ldots, S_n(I)]$, defined by $S_n(i) = 1$ if $s_n(i) = \sigma_i$, and $S_n(i) = 0$, otherwise. The jth column of C can be written as follows: $\left[f_j(\sigma_1), f_j(\sigma_2), \ldots, f_j(\sigma_I)\right]$, and the ith "row" of C is the value of $c[n]$, when $s_n = \sigma_i$.

For our demonstration starting case of a digital transition ($I = 4$), for a single noise source, ($M = 1$), we can use the simple notation, g_{ij}, where the ij indexes are the primary states, which "means": *from the i state to the j state*, where i and j take the value 0 (Low state) or 1 (High state). The digital transition signal is defined as a function $f: S \rightarrow C$ where $C \equiv \{00, 01, 11, 10\}^T$ (or $\{LL, LH, HH, HL\}^T$). Then, from s_n, we could set up a Transition Probability Matrix P of general term $P(i, j)$ ($=Pr\{s_{n+1} = s_j/s_n = s_i\}$). Then (for each noise source if necessary), will be assigned the relevant noise vector, here the simple one $\langle g_{00}(t), g_{01}(t), g_{10}(t), g_{11}(t)\rangle$, at t time.

The "superposition" of all associated parasitic waveforms (which can be stored in some library), g, coming from the commutating gate will give the final modulated signal. Moreover, $c[n]$ shall be modeled as a function of a Markov chain.

The very aim of our work is to compute, at any sensitive circuit node of interest, the spectral power density, called $S_y(f)$, which will be represented as a matrix; in fact, all the development of the algorithms are based on the matricial formalism, which appears, here, concise, and very general.

Dealing with the interference noise, means to develop a second-order analysis of the process, assuming it is a two order stationary one; In practice, we ought to calculate the correlation matrix function $R_x(k)$ (associated to the x noise source) of the vector $g(t, c)$, i.e.,

$$R_x[k] = E\left\{g(t, c[n])^* \cdot g(t, c[n+k])\right\}$$

$$(\text{*: transpose conjugate, for the general case}). \tag{5.127}$$

Let us consider a single cell (each cell will be further considered as an ideal current source or sink, which can be injected or absorbed—in power/

ground/substrate/package/board, which will be supposed having a linear electric behavior).

The digital signal *s* state can be defined as following (Figure 5.6):

- $L \rightarrow H$ (or: $0 \rightarrow 1$): with a probability = α
- $H \rightarrow L$ (or: $1 \rightarrow 0$): with a probability = β
  (*L* and *H* mean, respectively, high and low).

Then, Figure 5.7 represents the state diagram of such a "machine".

Let us suppose the "input" *i* is itself a Markov chain with a transfer probability matrix (TPM):

$$\pi = \begin{bmatrix} 1-\alpha & \alpha \\ \beta & 1-\beta \end{bmatrix} = \left(\begin{bmatrix} \pi_{00} & \pi_{01} \\ \pi_{10} & \pi_{11} \end{bmatrix} \right) \tag{5.128}$$

with: $\sum_j \pi_{ij} = 1$, i (=0 or 1): row number; j (=0 or 1): column number.

$$\pi_{01} = Pr\{j=1, \text{ if } i=0\} = \alpha \text{ and } \pi_{10} = Pr\{j=0, \text{ if } i=1\} = \beta$$

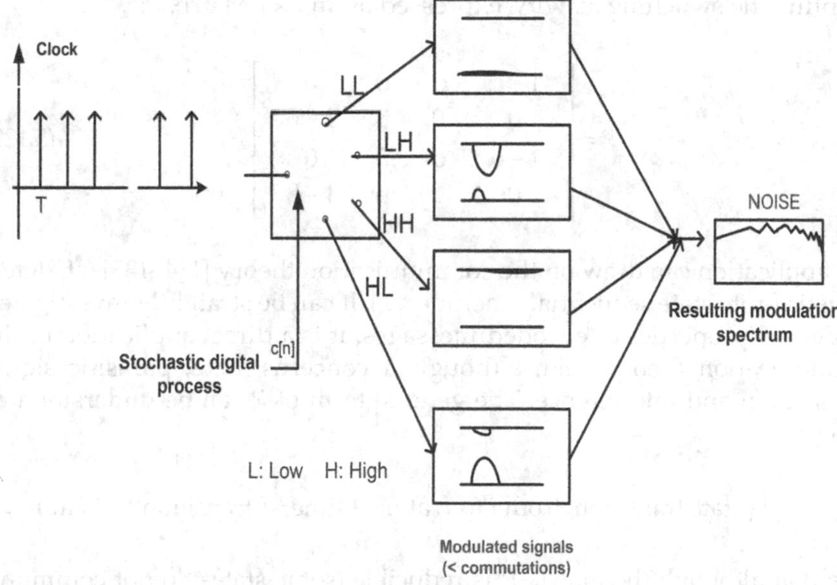

FIGURE 5.6
Block diagram of the parasitic signal propagation chain.

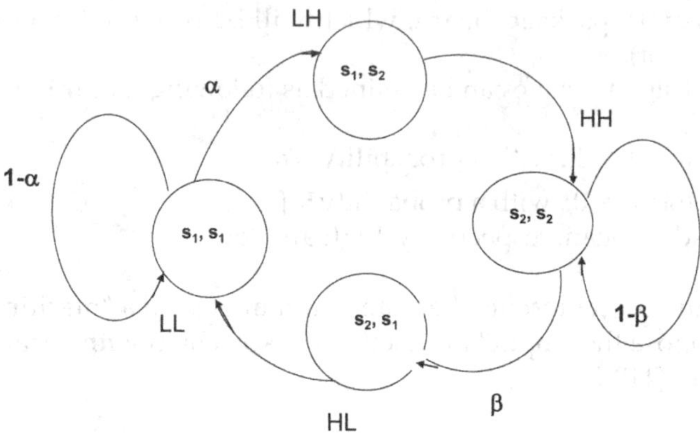

FIGURE 5.7
State diagram of the TPM.

Then:

$$\pi_{00} = Pr\{j=0, \text{ if } i=0\} = 1-\alpha, \text{ and } \pi_{11} = Pr\{j=1, \text{ if } i=1\} = 1-\beta.$$

A simple digital signal that may switch from *low* to *high* with probability α, and *low* to *high* with probability β at every clock tick, is modeled by four states (see Figure 5.7); it has a state Transition Probability Matrix (TPM) representing the switching activity, expressed as an $I \times I$ Matrix P:

$$P = \begin{bmatrix} 1-\alpha & \alpha & 0 & 0 \\ 0 & 0 & \beta & 1-\beta \\ 1-\alpha & \alpha & 0 & 0 \\ 0 & 0 & \beta & 1-\beta \end{bmatrix} \tag{5.129}$$

Our application can draw on the communication theory [134–137] (cf. deterministic finite state sequential machine, …). It can be straightly investigated as spectral properties of encoded messages. It is a direct application of the communication theory field, although it concerns, here, parasitic signal propagation and interference. The general term of P can be understood as follows:

$$P_{ij/kl} = Pr\{\text{state transition, from } i \text{ to } j, \text{ at } n+1 \text{ time/if transition } k \text{ to } l \text{ at } n\},$$

Note that although the matrix P is reducible (some states do not communicate with some another), it does not involve any drawback for the development of the rhythm.

A cumbersome, but not so difficult way to resolve this problem is depicted as follows. As we said, the aim is to obtain the spectral density generated by the digital block switches; that means, to find the Fourier transform of the corresponding autocorrelation function (cf. second-order moment).

This spectral density $w(f)$ can be split in two terms:

$$w(f) = w_d(f) + w_{\text{cont}}(f), \text{ where } f \text{ is the frequency.}$$

The discrete (W_d) component exhibits spectral lines or jumps at the frequency of the clock and its multiples ($f = n/T$), while the continuous part (w_{cont}) can be straightly derived by setting $z = \exp(j*2\pi f)$ into the z-transform of its correlation function:

$$w(z) = \sum_{k=-\infty}^{\infty} R_{\text{cont}}[k] z^{-k} \quad \left(R_{\text{cont}} \text{ is explicated hereafter}\right) \tag{5.130}$$

Moreover, the discrete random process S can be characterized by an autocorrelation matrix:

$$R(k) \equiv E\left\{S_n^{\,T}(j) S_{n+k}(i)\right\} = Pr\left\{s_{n+k} = \sigma_j, s_i = \sigma_n\right\} \; \left(T\text{: transpose}\right).$$

By using the matrix algebra tool, the whole correlation matrix (of the source codes $c[n]$) can be derived as follows:

$$R_c(k) = C^T R(k) C,$$

where $R(k)$ is the joint probability matrix between the random vectors S_n and S_{n+k}, with the "code words" $c[n] = S_n C$, where C is a matrix whom each column assigns a specific parasitic waveform. In a general point of view, a source code $b[n]$ is encoded into a sequence of code words $c[n]$; the "message" $c[n]$, is generally a nonlinear transformation, namely $c[n] = f(b[n])$, and can take the general form (input–output relation): $c[n] = g_0 \cdot b_n + g_1 \cdot b_{n-1} + \cdots + g_N \cdot b_{n-N}$; with $b[n] = (b_n, b_{n-1}, \ldots, b_{n-N})$.

Note, in our simple starting example, $N = 4$, with (see above):

$$g_0 \equiv g_{00}, g_1 \equiv g_{01}, g_2 \equiv g_{10}, g_3 \equiv g_{11}.$$

Finally, we explicit the whole correlation matrix as follows:

$$R[k] = \begin{cases} \Delta P^k & k \geq 0 \\ \left(P^T\right)^{-k} \Delta & k < 0 \end{cases} \tag{5.131}$$

P^k represents the k-step TPM, (note that, for $k = 0$, $R(0) = \Delta$, and $E\{S_n(j)S_n(i)\} = p_i \cdot \delta_{ij}$, where δ is the Kronecker symbol), with $\Delta = \text{diag}\{p(1), p(2), \ldots, p(I)\}$, which can be noted

$$\Delta = \begin{bmatrix} \infty \\ p_j \delta_{ij} \\ \end{bmatrix} \tag{5.132}$$

The elements of this diagonal matrix are the stationary (cf. the infinite symbol) vector components, which are derived from the equation:

$$p = pP\left(\text{with } p = \left[p_1, p_2, p_3, p_4\right]\right) \tag{5.133}$$

$R[k]$ can be as well decomposed in two "orthogonal" parts (cf. Vold's theorem [138]), $R[k] = R_{\text{cont}}[k] + R_d[k]$, where:

$$R_{\text{cont}}[k] = \begin{cases} \Delta H^k & k > 0 \\ \left(H^T\right)^{-k} \Delta & k < 0, \\ \Delta - \sum_{r=0}^{h-1} K_r & k = 0 \end{cases} \tag{5.134}$$

and a periodic component (cf. pure tones) with period h, given by

$$R_d[k] = \sum_{r=0}^{h-1} K_r \cdot \exp(j2\pi rk/h), \tag{5.135}$$

which is cyclic when $h > 1$.

We treat the case of arbitrary h, but, without loss of generality, we do not represent explicitly the global matrix, with $h > 1$, which is itself constructed via a cyclic form of the master matrix P. We can put in evidence the discrete and continuous part of these matrices, by transforming them, by suitable permutations of row (or/and columns), so that only zeros appear in the main diagonal.

Here K_r (equation 5.11) is the amplitude of the jumps (in our example, induced by a clock), and we have:

$$K_r \equiv J_r{}^* \, p^T p \, J_r \tag{5.136}$$

(asterisks denotes the transpose conjugate of the matrix, the hermitian one, for the general case)

Handling the matrix tool, K_r is directly related to the eigenvalues of modulus equal to one of the TPM (p, the stationary probability vector of the TPM, is an eigenvector of P—cf. equation 5.10.

In our problem, the solution of equation 5.10) gives the vector [α*(1 – β)/(α + β), (α*β)/(α + β), (β*α)/(α + β), β*(1 – α)/(α + β)] as solution.

The matrix H is then defined as follows:

$$H \equiv P - \sum_{r=0}^{h-1} \lambda_r J_r^* w p J_r, \tag{5.137}$$

where $w = [111\ldots11]'$: column vector.

That implies $H = P - P_\infty$ (P_∞ is the limit of P_k as $k \to \infty$)

As H has all its eigenvalues modulus <1 (by construction: P is split in two parts, H being the continuous one); therefore $R_{\text{cont}}(k)$ is absolutely summable.

Here J_r has the general form:

$$J_r = \begin{bmatrix} \lambda_r^{-1} I_1 & 0 & -- & 0 \\ 0 & \lambda_r^{-2} I_1 & -- & 0 \\ ----- & ----- & --- & --- \\ 0 & 0 & -- & \lambda_r^{-h} I_h \end{bmatrix} \tag{5.138}$$

$\lambda_r = \exp(j2pr/h)$, with $r = 0, 1, \text{h} - 1$, has a modulus equal to one.

In our example, there is only an eigenvalue of modulus one: $\lambda = 1$ (the others are 0—of order 2-, and 2-α – β); in that case:

$$R_d[k] = K_r. \tag{5.139}$$

We substitute (5.11) into (5.7) and taking into account that the limit of the matrix sum $M_i = \sum_{k=1}^{i} H^k z^{-k}$ converges to the matrix $M_\infty = zI - H$ (this result is easy if we use the so-called matrix formula: $\sum_{k=0}^{\infty} A^k = (I - A)^{-1}$, for any matrix A, provided all its eigenvalues have a modulus less than the unity). The eigenvalues λ_h of H are the value null, or non-null with a modulus lesser than 1 ($\lambda_h = 0, 0, 2\text{-}\alpha - \beta$).

After some cumbersome, bur relatively easy calculations, the continuous part of $w(z)$ (cf. equation 5.7) can be written as follows:

$$w(z) = \left(z^{-1}I - P^T\right)^{-1} C\left(zI - P\right)^{-1}, \text{with } C = \Delta - P^T \Delta P \tag{5.140}$$

with $z = \exp(j{*}2\pi f)$ and I is the identity matrix.

As we have been saying above, the spectral density can be considered as the sum of two parts: the continuous one, and the digital one. The spectral density of the modulated signal $x(t)$ can be formally written as follows:

$$S_x(f) = S_x^{\text{cont}}(f) + \sum_k \delta\left(f - f_k^d\right) \cdot S_{x,k}^d \tag{5.141}$$

The continuous spectral density of the modulated signal $x(t)$ is given by

$$S_x^{\text{cont}}(f) = \frac{1}{T} \cdot G^*(f) \cdot W_x\left(e^{j2\pi fT}\right) \cdot G^T(f) \tag{5.142}$$

where $G(f)$ is the Fourier transform of $g(t)$ and W_x (cf. equation 5.17) is the Fourier transform de $R_x(\tau)$, with:

$$R_x(\tau) = \frac{1}{T} \cdot \sum_{k=-\infty}^{+\infty} \int_{-\infty}^{+\infty} g^*(t) R(k) g(t + \tau - kT)\, dt \tag{5.143}$$

Indeed, note that $R_x(\tau)$ depends on the statistics of the Markov chain s_n, depending on $R(k)$.

On another hand, the noise signal $x(t)$ exhibits spectral lines (jumps) at frequency multiples of $f_0 = 1/T$, of amplitude

$$w_x^d(r) = C^* K_r C,\ r = 0, 1, \ldots, h-1 \tag{5.144}$$

Then $w_x^d(r)$, with K_r defined by (5.13) becomes

$$w_x^d(r) = \left| \frac{1}{T} p J_r G_r(f_0) \right|^2 \tag{5.145}$$

(G is the Fourier transform of the matrix $g(t)$).

The TPM P has only one eigenvalue, λ_0, equal to 1; all eigenvalues of modulus 1 result in Dirac distributions. In our case, only λ_0 results in a train of Dirac distributions at the frequency $1/T$ and its harmonics.

Our problem is analog to some modulation in digital communication.

Then $g(t; c[n])$ can be straightforwardly written $c[n].f(t)$, $f(t)$ being the parasitic basic waveform vector. Finally, $g(t) \equiv f(t) \cdot C$.

Therefore, the continuous part of the spectrum becomes

$$w_x^{\text{cont}}(f) = \frac{1}{T} \cdot |F(f)|^2 \cdot Wc(e^{j2\pi fT}), \text{ with } w_c(e^{j2\pi fT}) = C^* w(e^{j2\pi fT}) C \tag{5.146}$$

The discrete one becomes

$$S_x(f) = \frac{1}{T^2} \cdot \left|F(f)\right|^2 w_c^d \tag{5.147}$$

If now, we actually study the propagation of the noise, it necessary to take into account the transfer function of the "propagation channel", said *T*(*f*) (assuming linearity of the network).

We get the spectrum noise density, taking into account the network parasitic function transfer, *T*. The final power spectrum will be:

$$S_y(f) = T(f) \cdot S_x(f) \cdot T^*(f), \tag{5.148}$$

with:

$S_x(f)$: Power spectral density of the noise sources obtained using modulation theory

$T(j2\pi f)$: (generally) multi input (noise sources), multi output linear (number of sensitive points) system.

5.3.2.1 Numerical Experiences and Discussion

As a check to illustrate the algorithm explicated above, we study the case of signals injected into the analog power grid by one switching digital cell. Power (and ground) lines have to be efficiently modeled to predict voltage drops. Thanks to the ICEM (Integrated Circuit Emission Model) model [27,28], predict conducted-mode emission levels on a chip as well as in the application board is possible by means of SPICE-based analog simulations. Each part is modeled by an *RLC* network (see the schematic on Figure 5.8). We can derive very easily the following relation:

$$V_{\text{core}} = V_{\text{supply}} - \frac{L_{\text{sup}} \cdot j\omega + R_{\text{sup}}}{R_{\text{sup}} C_{\text{core}} \cdot j\omega - L_{\text{sup}} C_{\text{core}} \omega^2 + 1} \cdot I_{\text{act}}(j\omega) \tag{5.149}$$

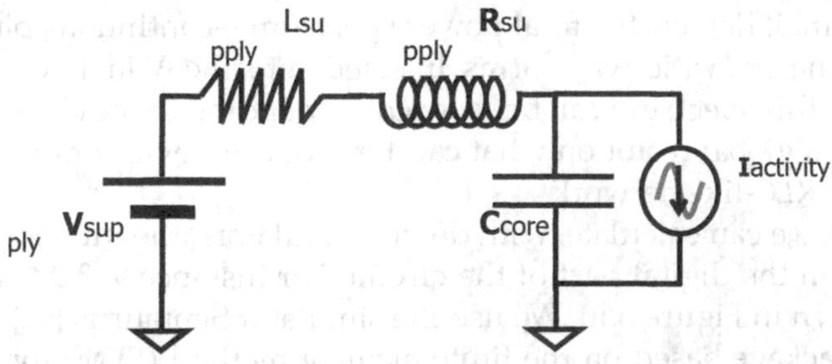

FIGURE 5.8
A *RLC*-like model for a supply network.

(or, in other words, the transfer function, T: $T^{-1} = jC\omega + (R + jL\omega)^{-1}$)

The values used for the parasitic components that model the power grid are based on measurements and extractions. Typically, we put: $R = 0.5\ \Omega$, $L = 0.2$ nH, C = 0.2 nF.

The resonant frequency is ten $f_0 = 8 \times 10^{18}$ Hz.

The core is represented by its transient consumption of current and by its capacitance. This dynamical consumption is the sum of all currents "demanded" by the logical gates of the digital circuit during its operating. The core capacitance is also the sum of all capacitances of digital gates and interconnects. For a single gate, for instance, its capacitance is made of its drain-substrate capacitance and its output capacitance itself made of interconnects and inputs of other gates. In this approach, it means that we consider the contribution to the supply lines voltage drop of each gate will be the same; in other words, the power bounces due to PCB, packaging, bonding wire, and other parasitic elements are more significant than the ones due to the power grid distribution. The power grid is considered as a unique voltage source. The power fluctuations do not depend of the topography of the design. In our case, and for the future developments, for a digital block, the switching activity current will be applied to one point: the middle point of this block.

As an example to validate the methodology and provide insight into some aspect of the problem, we consider a gate with only the state transition *LH* and *HL* (since the *LL* and *HH* state transitions involve a very low parasitic signal); the *LH* and *HL* parasitic signal shape will be assumed to be Gaussian versus the time (it is not so far from the reality, and considering moments greater than 2, for the signal shape, is not useful, for the moment, to validate our algorithm; what more is, a parasitic signal library is the best solution, for an industrial circuit simulation). We extracted the parameter mean values (hereafter), from a 90 nm CMOS process (see the beginning of this section):

LH: delay: 20 ns; standard deviation: 0.001 ns. $V_{peak} = 0.5$ V.
HL: delay: 60 ns; standard deviation: 0.01 ns. $V_{peak} = -0.15$ V.

Figure 5.9 shows the first simulations coming from the Markovian stochastic algorithm. It depicts the total power spectrum—continuous plus discrete parts—of the parasitic waveforms injected into the Vdd line. A periodic behavior in this spectrum can be seen; corresponding to the clock frequency. It indicates a global result on what can be produced by a commutating gate through an *RLC*-like network.

This way, we can calculate, with device simulators, the influence of spikes coming from the digital part of the circuit. For instance; a 3D CMOS structure is shown in Figure 5.10. We use the simulator Sentaurus [92]; which is a software package based on the finite element method (FEM) for the device study and the SPICE-based model for the CMOS devices is connected to the bulk regions. The TSV has a cylindrical shape filled with copper and with

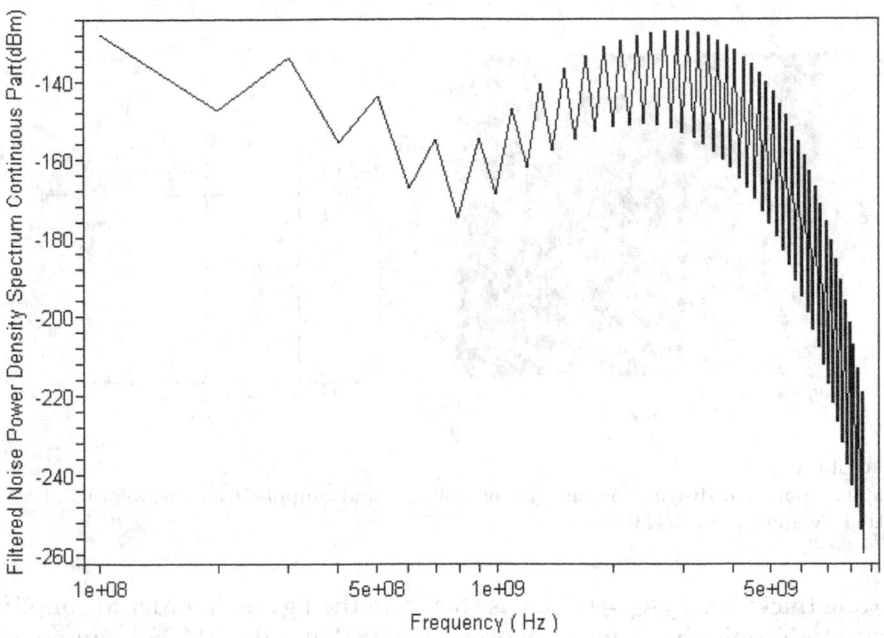

FIGURE 5.9
A typical calculated global noise spectrum of a gate commutating through a supply network.

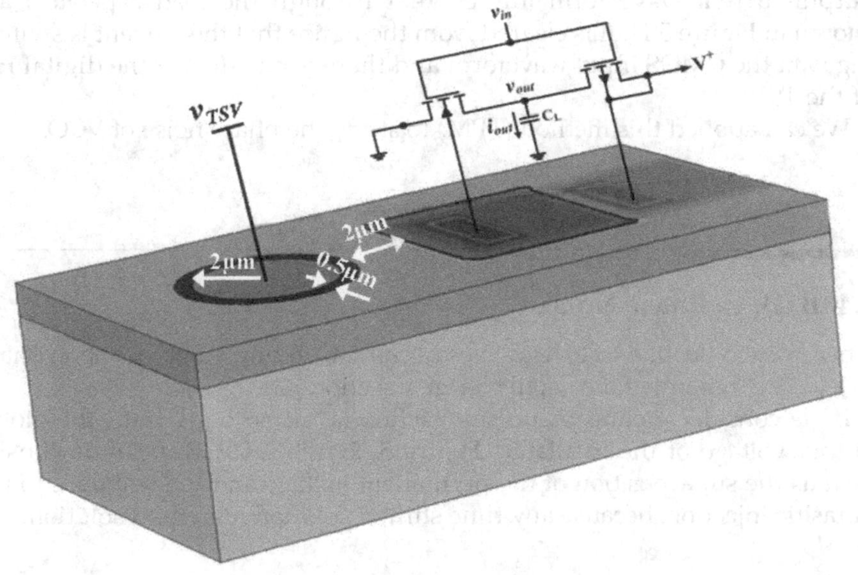

FIGURE 5.10
3D cross-section of TSV-CMOS mixed-mode coupling.

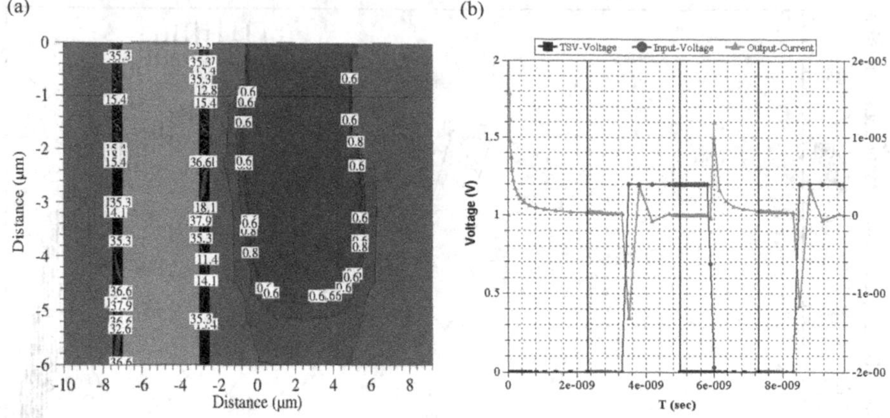

FIGURE 5.11
(a) The potential distribution and (b) the voltages and output-current waveforms at v_{in} = 0.0/1.2 V and v_{TSV} = 0.0/42.0 V.

oxide thickness T_{OXTSV} = 0.5 µm as shown in the figure. In order to simplify the study, only the coupling between the TSV and the nMOS is supposed. The nMOS transistor is placed near the TSV as illustrated in the figure. A Y-cut cross-section potential distribution is shown in Figure 5.10, for v_{in} = 0.0/1.2V and v_{TSV} = 0.0/42.0 V.

The waveforms of the TSV, CMOS input and output voltages and the output-current wave form (the current through the load capacitor) are shown in Figure 5.11. It is cleared from the figure that the current is switching w.r.t the CMOS input waveform and there is no effect of the digital HV of the TSV.

We can applied this method (TPM) to study the phase noise of VCO.

5.4 Back to Phase Noise

Now, we would like to extend the discussion to any value of the ϕ phase ($\phi \in [0, 2\pi]$), not only for a small charge injection [41].

If we consider a parasitic pulse injection (a "dirac") at τ time, the global output voltage of the oscillator (Figure 5.12: a SiGeC HBT) can be considered as the superposition of the permanent voltage and the response to the parasitic injection, because any time shifted solution remains a solution:

$$V_T(t) = V_s(t) + p(t) = V_0 \cdot \cos(\omega_0 t) + \frac{q}{C} \cdot \cos(\omega_0 t) \cdot u(t - t_0) \quad (5.150)$$

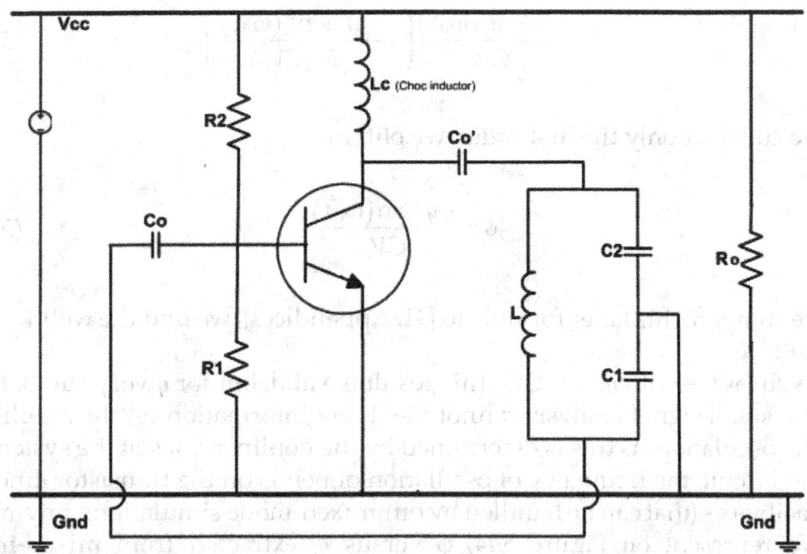

FIGURE 5.12
Schematic circuit of an *LC* oscillator.

This equation can be reformulated as follows:

$$V_T(t) = V_{T0}\cos[\omega_0\tau + \phi] \tag{5.150bis}$$

with

$$V_{T0} = \sqrt{V_0^{\,2} + \left(\frac{q}{C}\right)^2 + \frac{2qV_0\cos(\omega_0\tau)}{C}} \tag{5.151}$$

and

$$\phi = \arctan\left(\frac{-q\cdot\sin(\omega_0\tau)}{CV_0 + q\cdot\cos(\omega_0\tau)}\right) \pm k\pi \tag{5.151bis}$$

note: we get directly the ISF, by deriving (5.151bis):

$$\frac{d\Phi(\tau)}{d\tau} = \frac{-q\cdot\left(CV_0\cos(\omega_0\tau) + q\right)}{C^2V_0^{\,2} + q\cdot\left(q + 2CV_0\cdot\cos(\omega_0\tau)\right)} \tag{5.152}$$

Now, from these general formulas, and assuming a limited expansion of ϕ, we obtain

$$\phi = \frac{-q \cdot \sin(\omega_0 \tau)}{CV_0}\left(1 - \frac{q \cdot \sin^2(\omega_0 \tau)}{3 \cdot (CV_0)^2}\right) \tag{5.153}$$

If we consider only the first order, we obtain

$$\phi = \frac{-q \cdot \sin(\omega_0 \tau)}{CV_0} \tag{5.154}$$

If we compare this latter formula to [41, Appendices], we find the well-known result:

$\Gamma \sim \sin(\omega_0 t)$—with $q_{\max} = CV_0$; this result is valid, but for q very small. Note that a small signal analysis cannot yield any information on the amplitude of the oscillation, as this is determined by the nonlinearities of the system. In a real circuit, the frequency of oscillation depends on the transistor junction capacitances (that can be handled by our mixed-mode simulations, anymore).

We represent on Figure 5.14, ϕ versus q, extracted from mixed-mode simulations, and compare it to the "arctan" analytical solution of equation 5.151bis, where C is given by $C = \frac{C_1 \cdot C_2}{C_1 + C_2}$. In this typical case, we have from the "mixed-mode" $CV_0 = 100$ fc, in very good accordance with the "arctan" fit

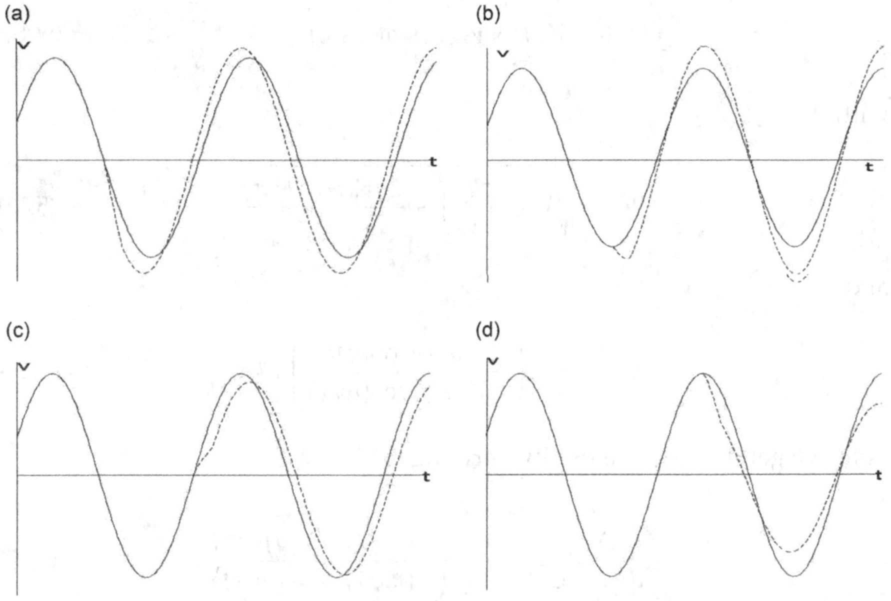

FIGURE 5.13
Mixed-mode simulation for Dirac pulses injected at (a) the zero-crossing when falling, (b) the bottom, (c) the zero-crossing when rising, and (d) the peak.

of equation 5.151bis, considering also the asymmetry (shift) along the y axis. We have chosen for this characteristic calculation (cf. Figure 5.13) an angle very near of –90°, "high amplitude" (not exactly 90°, to "observe" the influence of the denominator of equation 5.151bis, i.e., –81° (90°–10% of 90°).

Up to our knowledge, it is the first time that such comparisons are presented. Their interest is twofold: "microscopic" simulations seem validated by an analytical formulation, and we do not need to stay in the linear domain (see, for instance, [41]).

According to equations 5.151 and 5.151bis, phase or/and amplitude amplitudes can change after the perturbation. We represent on Figure 5.13 mixed-mode simulations when applying a pulse on Vdd (the choke inductance was replaced by a resistor). Our observations are as follows:

- ***At zero-crossing***

  For $\phi = \pm\phi_{max} = \arctan(\pm q/CV_0)$ $(\omega_0\tau = \pm\pi/2$, or $\tau = T/4$ or $3T/4)$:

  For zero-crossing falling $(\tau = T/4)$: $\phi = -\phi_{max}$; For zero-crossing rising $(\tau = 3T/4)$: $\phi = \phi_{max}$

$$\Delta V = \sqrt{V_0^2 + \left(\frac{q}{C}\right)^2} - V_0 \sim \frac{q^2}{2C^2V_0} \tag{5.155}$$

  This is quite negligible if q is small $(q/C \ll V_0)$.

- ***At the peaks***

  For $\phi = \phi_{min} = 0$: $\omega_0\tau = \pi$ or π, or $\tau = T/2$ or T):

$$\Delta V = \sqrt{\left(V_0 + \frac{q}{C}\right)^2} - V_0 \text{ or } \Delta V = \sqrt{\left(V_0 - \frac{q}{C}\right)^2} - V_0$$

  That mean for

$$\omega_0\tau = 2\pi : \Delta V = +\frac{q}{C} \text{ and } \omega_0\tau = \pi : \Delta V = -\frac{q}{C} \tag{5.155bis}$$

The HBT came from a 0.35 µm BiCMOS technology, the same that for the VCO presented in the Chapter 2 of the book.

Our "raw" simulations—Figures 5.13 and 5.14 - (q < 0 - current pulse negative -) show these results (see Figure 5.14), with some good numerical accuracy; if we consider the relation $|\Delta V| = \frac{|q|}{C}$, with typically $\Delta V = 1$–5 mV, that implies that $|q| = 5$–15 fc, which is well in the linear domain of ϕ versus q (see Figure 5.14).

If we inject a courant impulse at collector level (Figure 5.12), the instantaneous voltage change ΔV is given by

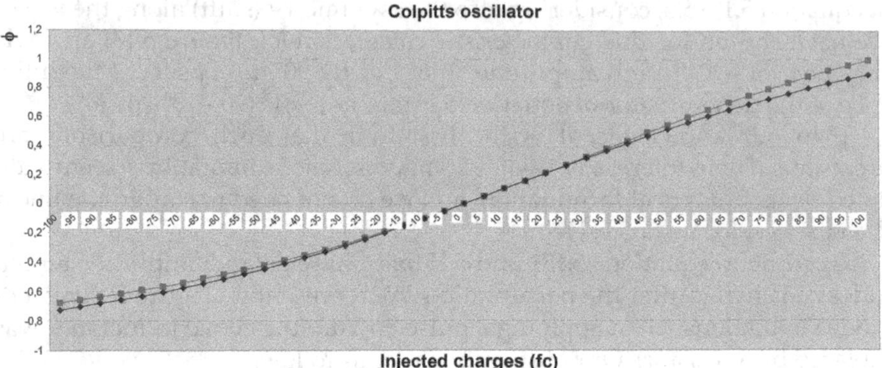

FIGURE 5.14
Phase shift versus injected charges (between collector and ground) for oscillator of Figure 5.12. Mixed mode (divJ/" Poisson"–"Kirchoff"): squares; "arctan" fit: diamonds (lower curve).

$$\Delta V = \frac{\Delta q}{C_{tot}},$$

where Δq is the total injected charge due to the current impulse (its integration over time) and C_{tot} is the total capacitive charge between the nodes of interest (amplifier output and ground).

If we "neglect" the resistance and if we represent the parasitic pulse current noise modeled as a "dirac", we get an equation of the following type:

$$\frac{1}{\omega_0{}^2}\frac{d^2V}{dt^2} + V = \frac{\Delta q}{C}\delta(t) \tag{5.156}$$

Note that the current pulse changes only the voltage across the capacitor.

By varying the delay of the disturbing Dirac impulse event (τ) during an oscillator period, we can access to the ISF function $\Gamma(\tau)$. In the first step of the work, we did not consider the magnitude shift as it can disappear with time; on another hand, the phase shift is preserved. The disturbance pulse can be a current spike in a capacitive node (injected charge), or a voltage spike in an inductive node. The well-known ISF function $\Gamma(\tau)$ is dimensionless and has the same period as the oscillator (or one can consider that $\Gamma(\omega_c\tau)$ has a 2π-period) [41]. Finally, we can write the excess phase to an Dirac impulse applied at t time:

$$h_\phi(t,\tau) = \frac{\Gamma(\omega_0\tau)}{q_{\max}}u(t-\tau) \tag{5.157}$$

where $u(t)$ is the unit step and q_{max} is the maximum charge across the capacitor C_{tot} ($q_{max} = C_{tot}$*$Vout_{max}$).

In other words, $\Gamma^{\phi}(\tau)$ function is a direct representation of the excess phase (phase shift), normalized by injected charge (for a current pulse).

If we inject a current impulse as shown in Figure 5.12, the instantaneous voltage change ΔV is given by

$$\Delta V = \frac{\Delta q}{C_{tot}},$$

where Δq is the total injected charge due to the current impulse (its integration over time) and C_{tot} is the total capacitive charge between the nodes of interest (amplifier output and ground).

If we "neglect" the resistance and if we represent the parasitic pulse current noise modeled as a "dirac", we get an equation of the following type:

$$\frac{1}{\omega_0{}^2}\frac{d^2V}{dt^2} + V = \frac{\Delta q}{C}\delta(t) \tag{5.158}$$

Note that the current pulse changes only the voltage across the capacitor.

By varying the delay of the disturbing Dirac impulse event (τ) during an oscillator period, we can access to the ISF function $\Gamma(\tau)$. In the first step of the work, we did not consider the magnitude shift as it can disappear with time; on another hand, the phase shift is preserved. The disturbance pulse can be a current spike in a capacitive node (injected charge), or a voltage spike in an inductive node. The well-known ISF function $\Gamma(\tau)$ is dimensionless and has the same period as the oscillator (or one can consider that $\Gamma(\omega_c\tau)$ has a 2π-period) [41]. Finally, we can write the excess phase to an Dirac impulse applied at t time:

$$h_{\phi}(t,\tau) = \frac{\Gamma(\omega_0\tau)}{q_{max}}u(t-\tau) \tag{5.159}$$

where $u(t)$ is the unit step and q_{max} is the maximum charge across the capacitor C_{tot} ($q_{max} = C_{tot}$*$Vout_{max}$).

In other words, $\Gamma^{\phi}(\tau)$ function is a direct representation of the excess phase (phase shift), normalized by injected charge (for a current pulse).

To go further, an interesting question is: how to calculate phase noise, when Dirac pulses are applying?

The $\frac{\phi}{\omega_0}(\tau)$ fluctuations are to the phase noise origin, depending on the noise sources (e.g., process induced traps for the HBT, jitter noise).

If we are interested by time fluctuations, we can study the following expression:

$$\tau_{\phi} = \frac{\phi(t+\tau) - \phi(t)}{\omega_0} \tag{5.160}$$

Taking the square of its RMS value:

$$\sigma^2 = \langle \tau_\phi{}^2 \rangle = \frac{\langle \phi(t+\tau) \rangle^2 - 2\langle \phi(t)\phi(t+\tau) \rangle + \langle \phi(t) \rangle^2}{\omega_0{}^2} \tag{5.161}$$

If now, we say that this process is a second-order stationary one, the RMS value does not depend on time, i.e.,

$$\phi_{RMS}{}^2 = \langle \phi(t)^2 \rangle = \langle \phi(t+\tau)^2 \rangle \tag{5.162}$$

The second term in the right side—equation 5.158—is the so-called correlation function: $C_\phi(\tau)$

Then equation 5.158 is reduced to:

$$\sigma^2 = \langle \tau_\phi{}^2 \rangle = 2\frac{\left(\phi_{RMS}{}^2 - C_\phi(\tau)\right)}{\omega_0{}^2} \tag{5.163}$$

Take back the parasitic waveform which comes from a capacitive charge injection, at ω_0 frequency, as treated above.

We have: $\phi = \frac{-q \cdot \sin(\omega_0 \tau)}{CV_0}$ (for weak values of ϕ; otherwise: take equation 5.153).

$$\phi_{RMS}{}^2 = \left\langle \left(\frac{-q \cdot \sin(\omega_0 \tau)}{CV_0} \right)^2 \right\rangle = \frac{q^2}{2C^2 V_0{}^2} \tag{5.164}$$

and

$$C_\phi(\tau) = \phi_{RMS}{}^2 \cdot \cos(\omega_0 \tau) \tag{5.164bis}$$

or

$$\sigma^2 = \frac{q^2}{C^2 V_0{}^2 \omega_0{}^2}\left(1 - \cos(\omega_0 \tau)\right), \tag{5.165}$$

which gives the associated noise power density:

$$S_\phi(\varpi) = \frac{q^2}{2.C^2 V_0{}^2 \omega_0{}^2}\left(2\delta(0) - \left[\delta(\omega - \omega_0) + \delta(\omega + \omega_0)\right]\right) \tag{5.166}$$

The last goal is to calculate the oscillator power spectral density output (PSD), via the autocorrelation output voltage:

$$S_V(\omega) = \mathrm{FT}\left(\frac{1}{t}\lim_{t\to\infty}\int_{-\frac{t}{2}}^{\frac{t}{2}}\left\langle V(t+\tau)\cdot V^*(t)\right\rangle dt\right) = \mathrm{FT}\left(V_0{}^2\lim_{t\to\infty}\int_{-\frac{t}{2}}^{\frac{t}{2}}\left\langle e^{j\omega_0\tau}e^{(\theta(t+\tau)-\theta(t))}\right\rangle dt\right) \tag{5.167}$$

It is easy to verify that the variance satisfy the relations:

$$\frac{d^2}{d\tau^2}\left[\sigma^2(\tau)\right] = C_\phi(\tau);\ \frac{d}{d\tau}\left[\sigma^2(0)\right] = 0;\ \sigma^2(0) = 0 \tag{5.168}$$

With these conditions and the fact that $\phi(t)$ varies slowly over time, equation 5.162 can be approximated as follows:

$$S_V(\omega) = \mathrm{FT}\left(V_0{}^2 e^{jw_0\tau} e^{\frac{\omega_0{}^2\cdot\sigma^2(\tau)}{2}}\right)\ (\text{FT: Fourier Transform}) \tag{5.169}$$

with σ^2 is given by (5.165)

Then, the equation 5.169 can be rewritten as follows:

$$S_V(\omega) = \mathrm{FT}\left(V_0{}^2\cdot\exp^{-\frac{q^2}{2C^2V_0{}^2}}\cdot\exp^{j\omega_0\tau}\cdot\exp^{\frac{q^2\cdot\cos(\omega_0\tau)}{2\cdot C^2V_0{}^2}}\right) \tag{5.170}$$

Find the Fourier Transform in equation 5.165 comes down to find the Fourier transform of

$$\exp^{\frac{q^2\cdot\cos(\omega_0\tau)}{2\cdot C^2V_0{}^2}},\ i.e.\text{: } S_V'(\omega) = 2\int_0^\infty \cos(\omega\tau)\exp^{\frac{q^2\cdot\cos(\omega_0\tau)}{2\cdot C^2V_0{}^2}}d\tau \tag{5.171}$$

This general formula (numerical solution), can be reduced, for the specific case where $q \ll CV_0$; equation 5.166 can be rewritten as follows:

$$S_V'(\omega) = \frac{2}{T}\sum_{n=1}^{\infty}\int_0^T \cos(n\omega_0\tau)\cdot\left(1+\frac{q^2\cdot\cos(\omega_0\tau)}{2\cdot C^2V_0{}^2}\right)d\tau \tag{5.172}$$

which can be reduced to:

$$S'_n\omega = \delta(0) + \frac{q^2}{2C^2 V_0^2}\left[\delta'(-\omega_0) + \delta'(\omega_0)\right] \tag{5.173}$$

At this stage, it seems important to present an original point of view of such oscillators: the virtual damping concept [139]; it is considered as "phase diffusion", as physical diffusion of particles, called Brownian motion [123]. Oscillator phase noise has been studied from an in-depth mathematical point of view [41].

In this case, the δ-function response is not adequate (the system is not linear or quasi-linear), but is broadened into a finite bandwidth. The fluctuations of the oscillator phase are quite analogous to Brownian motion of a free particle; hence the phase fluctuation can become quite large, without any cost in energy; the spectral linewidth which involves the phase fluctuations cannot be treated by quasi-linear methods.

The signal amplitude of the damped oscillator output takes the form [139]:

$$V_s(t) = V_0 \exp(-Dt)\cdot\cos(\omega_0 t + \phi) \tag{5.174}$$

(assuming a Gaussian white noise; then $\langle \phi^2 \rangle = 2Dt$);

where D can be called "diffusion frequency".

According [139], taking N (with N very large) identical oscillators, with the same initial condition (e.g., $\phi = 0$ at $t = 0$), each oscillator loses its initial phase information after a "sufficient longtime". [41] gives some values of phase noise ($\sim 2D/(\Delta\omega)^2$) for different damping: D^{-1} in the range [1 ms, 1 year], or larger…

We think that we will have the same result for a single oscillator, for a very long time (cf. ergodicity).

We postulate: any oscillator will be damped, and stopped, in one week, or one century, if there is no intrinsic noise opposed to this phenomenon. We can "recall" the Langevin though when he has found again the so-called formula of the Brownian diffusion coefficient of Einstein [123]: "from the Newton law, and introducing a friction force for the particle, it will stop… in a given time: but, it's a noise (that he supposed white, with a mean value null), which impeaches the slow down movement".

In "the real life", intrinsic white noises (with a null mean value) ought to exist into or near the circuits; their correlation function can be written as follows:

$$\langle \eta(t_1)\cdot\eta(t_2)\rangle = A^2\cdot\delta_{ij}\cdot\delta(t_1 - t_2) \tag{5.175}$$

Let now introduce a "parasitic" solution, x_p; the total solution of the harmonic oscillator

$$\frac{d^2V}{dt^2} + \omega_0{}^2 V = 0 \tag{5.176}$$

can be written:

$$V_T(t) = V_0 \cdot \cos(\omega_0 t + \phi) + k \cdot x_p \tag{5.177}$$

(where x_p can be called a Langevin noise source)

After some cumbersome but not so difficult calculations, the solution (we suppose constant the frequency…) for x_p, at the first order, has the following form:

$$V_s(t) = V_0 \cdot \cos(\omega_0 t) + \gamma\left[\left(\cos(3\omega_0 t) - \cos(\omega_0 t)\right) + 12\omega_0 t \cdot \sin(\omega_0 t)\right]. \tag{5.178}$$

In this solution, we meet a term proportional to time. The oscillator should diverge (in fact, the voltage amplitude should be limited by the supply).

We suggest that the damping response (cf. equation 5.174) could be compensated by the x_p term, explained in equation 5.177, especially for D^{-1} large enough.

5.5 Conclusion

Modeling 3D noise, even 2D, is, for the most part, a failure due to physical modeling problems: for example, discerning the hole and electron contribution to noise in devices. Master equations, like the Boltzmann equation, associated with second-order moments should be developed more for theoretical study of noise.

For submicronic devices, new effects, like nonstationary phenomena, crop up. By reducing the length of a device to some mean free paths, each carrier undergoes very few collisions on its path from one electrode to another. Another example is correlation of noise sources in space, which, are not independent, even though autonomous. Moreover, intensities of noise can increase during voltage ramp applied to a circuit. Other investigations must be made, in the case of the following points:

- Quantum phenomena: duration of carrier collisions versus energy exchange with phonons.
- Proposing rigorous techniques or methods of noise reduction, even if it is subtler than it seems in this ambiguous domain.

The program, to be developed in the future, could focus on the following main areas.

- Continuing basic studies on microscopic behavior of noise sources up to 3D, with theories and experiments, in order to have a better understanding of phenomena.
- Study submicronic, nanometric transistors, used for high-speed logic in information technology. The objective is to study the $1/f$ noise and the RTS (Random Telegraphic Signal) noise, in order to better understand ultimate (Bi-)CMOS, and develop compact electric models of device noise to correctly and fully simulate the behavior of these devices and the related circuits.
- High-speed III–V transistors (HEMT and HBT), as important elements of future high-frequency broadband telecommunication systems. Low-frequency analyses ($1/f$, G–R and RTS noise) and high-frequency noise will be conducted on semiconductor devices and material (bulk, epitaxial layers, flexible electronics).
- Noise in photoelectric detectors and lasers, particularly in multiquantum well structure, for high-speed, optoelectronics, and gigabit communications.
- Sensors, in particular those using magnetic effects.

We deal, also, with a concise and general method, giving direct expression of the discrete and continuous component of noise power spectra, involved by the gate commutations of complex circuits, at any node of them.

Closed-form expressions have been worked out, for the spectral lines and the continuous spectral density of commutation processes, considered as ergodic stationary Markov chains (MC). For the sake of generality, the algorithms essentially rely upon matrix analysis. As a first example, we have been studying the perturbation induced by a gate commutation, of which the "victim" is the supply. With this work, we will go on to define the way to design as easily as possible a low noise digital block, in some simple way. One of the problems in front of us will be to evaluate the effect of a great number (some millions…) of switching gates, on certain particularly sensitive point of the heterogeneous chip.

A final insight in time-depending phase noise analysis of integrated oscillators is involved. Future trends will lead to develop specific noise study, theoretically and with mixed-mode simulations, as jitter one, or induced by physical process steps or/and carrier traps, taking into account, without too simplistic hypothesis, the substrate influence. It should be noted that, thanks to the modeling algorithms embedded in the simulator, we should access to some intrinsic noise source influence. In others words, our method should not require specific "artificial" noise source to be introduced at some nodes of interest, to access in some detail to the phase noise.

6

General Conclusion

This book provides a global design method for the conception of 3D integrated circuits in electrical, thermal, electrothermal, and noise field models. Included are simulation methods for single TSV, TSV matrix, multilayers substrate, multiprocessors, and heat dissipation design.

We present some thermal and electrothermal modeling methods. We propose an analytical thermal solution for multilayers substrate nonstationary temperature distribution. In the 3D IC design, the thermal issues are crippling issues, especially when we scale down to nanoscale. The electrothermal simulator is totally compatible with SPICE (-like) simulators. This work constructs the electrothermal analysis framework for substrate, thermal connections (3D via), and system level 3D PDN. We provide an extensive study of power and thermal dissipation issues in 3D integration considering several aspects such as:

- Single and multilayer auto-heating effects on 3D power and thermal dissipation
- TSVs and substrate impact on power and thermal integrity and their design implications
- Heat sink's thermal conductivity's influence on the electrothermal effects

As a solution to the local heat dissipation, a passive heat dissipation component, a flat heat pipe (FHP), is proposed as a prospective device. The FEM-based model (COMSOL Multiphysics) is very accurate in the 2D and 3D simulations of heat pipe. Precise boundary conditions can be found; it considers the pressure field in the calculation, and especially it can extract the capillary pressure directly from the model, which will be very useful in the future study of wick layer's capillary effects.

We propose also noise analysis methods and investigate the keep-out-of-zone (KOZ) of TSV to transistor, such as nMOS and CNFET. The investigation methods of electrical noise impact and thermo-mechanical noise impact show large influence to nMOS like device (pMOS, pFET, and nFET). But for CNFET, due to the highly isolation between carbon nano-tube and substrate, the parasitic effect of TSV has no influence to it.

To realize the electrical, thermal, and electrothermal modeling of single TSV, TSV matrix and multilayers 3D IC clarify the coupling noise processes involved in multilayer's substrate and give the EDA design tools. But much remains to be done; a complex real case should be studied (cf. CMOS imager, NMR system, etc.)

Extension of this work will concern an M3D (3D Monolithic) electrothermal modeling, which can directly realize nanoscale device level design (a CMOS inverter can be obtained in nanoscale by connection of a pMOS and nMOS with a M3D). And more RF devices can be generated by using TSV or M3D technology in micro and nano scale; for example, micro RF antenna with TSV, TSV-based thermal coupler, TSV or M3D-based logic cell, and other 3D sensors.

Recently, the research of 3D IC has extended to optic-electrical design. Thus, one of the future research interests will be the EDA optic-electronic co-design. But this will be kind of useless, because in the optical application, the substrate and transmission medium are always glasses, and in this case, the parasitic effects are always negligible, in micro scale. But due to the high speed transmission signal, the optical circuits can be easily overheated. So the optic-electrical-thermal co-design will be one important and interesting work in the future.

References

1. E. P. Debenedictis, M. Badaroglu, A. Chen, T. M. Conte, and P. Gargini, "Sustaining Moore's law with 3D chips," *Computer (Long. Beach. Calif.)*, vol. 50, no. 8, pp. 69–73, 2017.
2. A. Todri-Sanial, "3D integration overview; Overview of physical design issues for 3D-integrated circuits," In *Physical Design for 3D Integrated Circuits*, Eds. A. Todri-Sanial and C. S. Tan, CRC Press, Boca Raton, November 24, 2015.
3. P. A. Gargini, "How to successfully overcome inflection points, or long live Moore's law," *Comput. Sci. Eng.*, vol. 19, no. 2, pp. 51–62, 2017.
4. R. J. Gutmann and J.-Q. Lu, "Wafer-level three-dimensional ICs for advanced CMOS integration," In *Design of 3D Integrated Circuits and Systems*, Ed. R. Sharma, CRC Press, Boca Raton, November 12, 2014.
5. AMD, "High-bandwidth memory (HBM) reinventing memory technology industry standards on-die GPUs," 2015. www.amd.com/.../High-Bandwidth-Memory-HBM.pdf.
6. K. Ashraf, S. Smith, and S. Salahuddin, "Electric field induced magnetic switching at room temperature: Switching speed, device scaling and switching energy," *Tech. Dig. - Int. Electron Devices Meet. IEDM*, vol. 2, pp. 601–604, 2012.
7. M. M. Shulaker, G. Hills, N. Patil, H. Wei, H. Y. Chen, H. S. P. Wong, and S. Mitra, "Carbon nanotube computer," *Nature*, vol. 501, no. 7468, pp. 526–530, 2013.
8. D. E. Nikonov and I. A. Young, "Benchmarking of beyond-CMOS exploratory devices for logic integrated circuits," *IEEE J. Explor. Solid-State Comput. Devices Circuits*, vol. 1, no. c, pp. 3–11, 2015.
9. P. Ramm, A. Klumpp, J. Weber, and M. M. V. Taklo, "3D system-on-chip technologies for more than Moore systems," *Microsyst. Technol.*, vol. 16, no. 7, pp. 1051–1055, 2010.
10. M. Motoyoshi and M. Koyanagi, "3D-LSI technology for image sensor," *J. Instrum.*, vol. 4, no. 3, pp. 2–14, 2009.
11. J. Charbonnier, D. Henry, F. Jacquet, B. Aventurier, G. Enyedi, N. Bouzaida, V. Lapras, and N. Sillon, "Wafer level packaging technology development for CMOS image sensors using through silicon vias M1 pad on front side," *IEEE Electronics System Integration Technology Conference*, pp. 141–148, 2008.
12. K. Chang, A. Koneru, K. Chakrabarty, and S. K. Lim, "Design automation and testing of monolithic 3D ICs : Opportunities, challenges, and solutions," *Proceedings of the 36th International Conference on Computer-Aided Design*, pp. 805–810, 2017.
13. S. Kannan, R. Agarwal, A. Bousquet, G. Aluri, and H. S. Chang, "Device performance analysis on 20 nm technology thin wafers in a 3D package," *IEEE International Reliability Physics Symposium Proceedings*, vol. 2015-May, pp. 4C41–4C45, 2015.

14. P. Batude, T. Ernst, J. Arcamone, G. Arndt, P. Coudrain, and P. E. Gaillardon, "3-D sequential integration: A key enabling technology for heterogeneous co-integration of new function with CMOS," *IEEE J. Emerg. Sel. Top. Circuits Syst.*, vol. 2, no. 4, pp. 714–722, 2012.
15. A. Topol, D. C. La Tulipe, L. Shi, S. M. Alam, D. J. Frank, S. E. Steen, J. Vichiconti, D. Posillico, M. Cobb, S. Medd, J. Patel, S. Goma, D. DiMilia, M. T. Robson, E. Duch, M. Farinelli, C. Wang, R. A. Conti, D. M. Canaperi, L. Deligianni, A. Kumar, K. T. Kwietniak, C. D'Emic, J. Ott, A. M. Young, K. W. Guarini, and M. Ieong, "Enabling SOI-based assembly technology for three-dimensional (3d) integrated circuits (ICs)," *IEEE International Devices Meeting 2005. IEDM Technical Digest*, pp. 352–355, 2005.
16. P. R. Philip Garrou and C. Bower, *Handbook of 3D Integration: Technology and Applications of 3D Integrated Circuits*, Wiley-VCH Verlag GmbH & Co. KGaA, Weinheim, 2008.
17. K. Chang, B. W. Ku, S. Sinha, and S. K. Lim, "Full-chip monolithic 3D IC design and power performance analysis with ASAP7 Library," *ICCAD '17 Proceedings of the 36th International Conference on Computer-Aided Design*, Irvine, California, November 13–16, 2017, IEEE Press Piscataway, NJ, USA, pp. 1005–1010, 2017.
18. P. Batude, M. Vinet, A. Pouydebasque, C. Le Royer, B. Previtali, C. Tabone, J. M. Hartmann, L. Sanchez, L. Baud, V. Carron, A. Toffoli, F. Allain, V. Mazzocchi, D. Lafond, S. Deleonibus, and O. Faynot, "3D monolithic integration," *Proceedings of IEEE International Symposium on Circuits and Systems*, pp. 2233–2236, 2011.
19. Y. Ma, O. Valorge, J. R. Cardenas-Valdez Aldez, F. Calmon, J.C. Nuñez–Perez, F. Calmon, J.C. Nuñez–Perez, J. Verdier, and C. Gontrand, "3-D interconnects with IC's stack global electrical context consideration," In *Noise Coupling in System-On-Chip*, Ed. T. Noulis, CRC Press, Boca Raton, 2018.
20. O. Valorge, F. Sun, J.-E. Lorival, F. Calmon, and C. Gontrandin, "Mixed-signal IC design addressed to substrate noise immunity in bulk silicon; toward 3D circuits," in *Mixed-signal circuits*, Ed. T. Noulis, CRC Press, Boca Raton, 2016.
21. V.F. Pavlidis and E. G. Friedman, *Three-Dimensional Integrated Circuit Design*, Elsevier Inc., Amsterdam, 2009.
22. T.-J. Brazil "The modeling and simulation of high-frequency electronic circuits," *Int. J. Circ. Theor. Appl.*, vol. 35, pp. 533–546, 2007.
23. J. Briaire and K. S. Krisch, "Principles of substrate crosstalk generation in CMOS circuits," *IEEE Trans. Comput.-Aided Des. Integr. Circuits Syst.*, vol. 19, no. 6, pp. 645–653, 2000.
24. M. Felder and J. Ganger, "Analysis of ground-bounce induced substrate noise coupling in a low resistive bulk epitaxial process: Design strategies to minimize noise effects on a mixed-signal chip. *IEEE Trans. Circ. Syst.*, vol. 46, no. 11, pp. 1427–1436, November 1999.
25. M. Baradoglu, G. Van der Plas, P. Wambacq, L. Balasubramanian, K. Tiri, I. Verbauwhede, S. Donnay, G.G.E. Gielen, and H.J. De Man, "Digital circuit capacitance and switching analysis for ground bounce in ICs with a high-ohmic substrate," *IEEE J. Solid State Circ.*, vol. 39, no. 7, pp. 1119–1130, 2004.
26. S. Reddy , R. Murgai, "Accurate substrate noise analysis based on library module characterization," *19th International Conference on VLSI Design held jointly with 5th International Conference on Embedded Systems Design (VLSID'06)*, pp. 3–7, January 2006.

27. E. Sicard, EMC modeling of integrated circuits usig IC-EMC - Archive ouverte, 2015. https://hal.archives-ouvertes.fr/hal-01225370/document.
28. O. Valorge, C. Andrei, B. Vrignon, F. Calmon, C. Gontrand, J. Verdier, and P. Dautriche, Using ICEM models for substrate noise characterization in mixed signal IC's. *Zurich International Symposium on Electromagnetic Compatibility*, pp. 353–356, February 2005, Zurich, Switzerland.
29. O. Valorge, C. Andrei, F. Calmon, J. Verdier, C. Gontrand, and P. Dautriche, "A simple way for substrate noise modeling in mixed-signal ICs," *IEEE Trans. Circuits Syst. I, Reg. Papers*, vol. 53, no. 10, pp. 2167–2177, October 2006.
30. Y. Rolain, W. Van Moer, G. Vanderstenn, and M. van Heijningen, "Measuring mixed-signal substrate coupling," *IEEE Trans. Instrum. Meas.*, vol. 50, pp. 959–964, 2001.
31. D. K. Su, M. J. Loinaz, S. Masui, and B. A. Wooley, "Experimental results and modeling techniques for substrate noise in mixed-signal integrated circuits," *IEEE J. Solid State Circ.*, vol. 28, no. 4, April 1993, pp. 420–430.
32. M. Xu, D. K. Su, D. K. Shaeffer, T. H. Lee, and B. A. Wooley, "Measuring and modeling the effects of substrate noise on the LNA of a CMOS GPS receiver," *IEEE J. Solid State Circ.*, vol. 36, no. 3, pp. 473–485, March 2001.
33. M. van Heijningen, M. Baradoglu, S. Donnay, G. G. E. Gielen, and H. J. De Man, "Substrate noise generation in complex digital systems: Efficient modeling and simulation methodology and experimental verification," *IEEE J. Solid State Circ.*, vol. 37, no. 8, pp. 1065–1072, August 2002.
34. O. Valorge, C. Andrei, F. Calmon, C. Gontrand, J. Verdier, and P. Dautriche, "Design slop constraint for reducing noise generation and coupling mechanisms in mixed signal IC's," *International Symposium on Industrial Electronics*, May 2004, Ajaccio, France.
35. C. Soens, G. Van der Plas, P. Wambacq, and S. Donnay, "Performance degradation of an LC-tank VCO by impact of digital switching noise," *European Solid-State Circuits Conference*, 2004, Leuven, Belgium.
36. C. Soens, C. Crunelle, P. Wambacq, G. Vandersteen, S. Donnay, Y. Rolain, M. Kuijk, and A. Barel, "Characterization of substrate noise impact on RF CMOS integrated circuits in lightly doped substrates," *IEEE Instrumentation and Measurement Technology Conference*, 2003.
37. M. Nagata, J. Nigai, K. Hijikata, T. Morie, and A. Iwata, "Physical design guides for substrate noise reduction in CMOS digital circuits," *IEEE J. Solid State Circ.*, vol. 36, no. 3, pp. 539–549, March 2001.
38. M. Baradoglu, M. van Heijningen, V. Gravot, J. Compiet, S. Donnay, G. G. E. Gielen, and H. J. De Man, "Methodology and experimental verification for substrate noise reduction in CMOS mixed-signal ICs with synchronous digital circuits," *IEEE J. Solid State Circ.*, vol. 37, no. 11, pp. 1383–1395, November 2002.
39. O. Valorge, C. Andrei, B. Vrignon, F. Calmon, C. Gontrand, J. Verdier, and P. Dautriche, "On a standard approach for substrate noise modeling in mixed signals IC's," *Proceedings of IEEE International Conference on Microelectronics*, December 2004.
40. S. B. Worm and N. van Dijk, "Susceptibility analysis of oscillators by means of the ISF-method," *EMC Compo 04 – International Workshop on Electromagnetic Compatibility of Integrated Circuits*, Angers 2004, France.
41. A. Hajimiri and T. H. Lee, "A general theory of phase noise in electrical oscillators," *IEEE J. Solid State Circ.*, vol. 33, no. 2, pp. 179–194, February 1998.

42. O. Valorge, F. Calmon, C. Andrei, C. Gontrand, and P. Dautriche, "Mixed-signal IC design to enhance subtrate noise immunity in bulk silicon technology," *Analog Integr. Circuits Signal Process.*, vol. 63, no. 2, pp. 185–196, 2010.
43. COMSOL Multiphysics, www.comsol.com.
44. J. Kanapka, J. Phillips, and J. White, "Fast method for extraction and sparsification of substrate coupling," *Proceedings of the 37th Design Automation Conference*, pp. 738–743, June 2000.
45. FastHenry and FastCap, www.fastfieldsolvers.com.
46. HFSS by Ansoft Corporation.
47. CadMOS *Seismic.*
48. F. Hertzel and B. Razavi, "A study of oscillators jitter due to supply and substrate noise," *IEEE Trans. Circuits Syst. II*, vol. 46, no. 1, pp. 56–62, January 1999.
49. F. Calmon, C. Andrei, O. Valorge, J. C. Nuñez Perez, J. Verdier, and C. Gontrand, "Impact of low-frequency substrate disturbances on a 4.5 GHz VCO," *Microelectron J.*, vol. 37, no. 10, pp. 1119–1127, 2006.
50. A. Hajimiri and T. H. Lee, *The Design of Low Noise Oscillators*, Kluwer Academic Publishers, Boston, 2001.
51. H. Jacquinot, J. Majos, and P. Senn, "5 GHz low-noise bipolar and CMOS monolithic VCOs," *European Solid-State Circuits Conference*, 2000, Stockholm, Sweden.
52. B. Razavi, "A study of phase noise in CMOS oscillators," *IEEE J. Solid State Circ.*, vol. 31, no. 3, pp. 331–343, March 1996.
53. H. Liao, S. C. Rustagi, J. Shi, and Y. Z. Xiong, "Characterization and modeling of the substrate noise and its impact on the phase noise of VCO," *IEEE Radio Frequency Integrated Circuits Symposium*, 2003.
54. C. Andrei, O. Valorge, F. Calmon, J. Verdier, and C. Gontrand, "Substrate noise impact on a 4.5 GHz VCO for different inductor shield structures," *Proceedings of Analog VLSI workshop*, October 2004, Bordeaux, France.
55. T. Yoshinaga and M. Nomura, "Trends in R&D in TSV technology for 3D LSI packaging," Science and Technology Trends, quarterly review, no. 37, October 2010.
56. Z. Xu and J.-Q Lu, "High-speed design and broadband modeling of through-strata-vias (TSVs) in 3D integration," *IEEE Trans. Compon. Packag. Manuf. Technol.*, vol. 1, no. 2, p. 15162, February 2011.
57. A. Todri, S. Kundu, P. Girard, A. Bosio, L. Dilillo, and A. Virazel, "A study of tapered 3-D TSVs for power and thermal integrity," *IEEE Trans. Very Large Scale Integr. Syst.*, vol. 21, no. 2, pp. 306–319, 2013.
58. Y. Liang and Y. Li, "Closed-form expressions for the resistance and the inductance of different profiles of through-silicon vias," *IEEE Electron Device Lett.*, vol. 32, no. 3, pp. 393–395, March 2011.
59. I. Savidis and E. G. Friedman, "Closed-form expressions of 3-D via resistance, inductance and capacitance," *IEEE Trans. Electron Devices*, vol.56, no. 9, pp. 1873–1881, September 2009.
60. D. H. Kim, S. Mukhopadhyay, and S. K. Lim, "Fast and accurate modeling of through-silicon-via capacitive coupling," *IEEE Trans. Compon. Packag. Manuf. Technol.*, vol. 1, no. 2, pp. 168–180, February 2011.
61. Y. Eo and W. R. Eisenstadt, "High-speed VLSI interconnect modeling based on S-parameter measurements," *IEEE Trans. Compon., Hybrids, Manuf. Technol.*, vol. 16, no. 5, pp. 555–562, August 1993.

62. A. Weisshaar, H. Lan, and A. Luoh, "Accurate closed-form expressions for the frequency-dependent line parameters of on-chip interconnects on lossy silicon substrate," *IEEE Trans. Adv. Packag.*, vol. 25, no. 2, pp. 288–296, May 2002.
63. MATLAB, http://mathworks.com.
64. Advanced Design System - ADS, www.home.agilent.com.
65. C. Xu, V. Kourkoulos, R. Suaya, and K. Banerjee, "A fully analytical model for the series impedance of through-silicon vias with consideration of substrate effects and coupling with horizontal interconnects," *IEEE Trans. Electron Devices*, vol. 58, no. 10, pp. 3529–3540, October 2011.
66. E. B. Rosa, "The Self and Mutual Inductances of Linear Conductors," U. S Bulletin of Standards, 1907.
67. J. Nurmi, H. Tenhunen, J. Isoaho, and A. Jantsch, *Interconnect-Centric Design for Advanced SoC and NoC*, Kluwer Academic Publisher, Boston, 453 pages, 2004.
68. Y. Ma, O. Valorge, J. R. Cardenas-Valdez, J. C. Nuñez–Perez, J. Verdier, F. Calmon, C. Gontrand, "Electro-thermal considerations dedicated to 3-D integration; noise coupling," In *Noise couplings in systems on chip*, Ed. T. Noulis, CRC Press, Boca Raton, 2018.
69. Y. Ma, F. Calmon, and C. Gontrand, "Electric compact model of TSV and an associated developed tool," *6th Electronic System-Integration Technology Conference*, ESTC, 2016.
70. M. Stucchi, J. De Vos, A. Jourdain, Y. Li, G. Van Der Plas, K. Croes, and E. Beyne, "Anomalous C-V inversion in TSVs: The problem and its cure," *IEEE Trans. Electron Devices*, vol. 65, no. 4, pp. 1473–1479, 2018.
71. O. Valorge, F. Sun, J.-E. Lorival, F. Calmon, and C. Gontrand, "Mixed-signal IC design addressed to substrate noise immunity in bulk silicon; chap, 50 p, décembre 2015 towards 3D circuits," In *Mixed Signal Circuits*, Taylor & Francis Group, Boca Raton, 2017.
72. C. Virtuoso, Spectre Circuit Simulator, www.cadence.com.
73. L.-B. Qian, Z.-M. Zhu, and Y.-T. Yang, "Three-dimensional global interconnect based on a design window," *Chinese Phys. B*, vol. 20, no. 10, p. 108401, 2011.
74. D. C. Sekar, A. Naeemi, R. Sarvari, J. A. Davis, and J. D. Meindl, "IntSim: A CAD tool for optimization of multilevel interconnect networks," *IEEE/ACM International Conference on Computer-Aided Design. Digest of Technical Papers ICCAD*, no. 1, pp. 560–567, 2007.
75. P. Dixit, H. Viljanen, J. Salonen, T. Suni, J. Molarius, and P. Monnoyer, "Fabrication, electrical characterization and reliability study of partially electroplated tapered copper through-silicon vias," *Proc. Tech. Pap. - Int. Microsystems, Packag. Assem. Circuits Technol. Conf. IMPACT*, pp. 190–193, 2013.
76. C. Xu, V. Kourkoulos, R. Suaya, and K. Banerjee, "A fully analytical model for the series impedance of through-silicon vias with consideration of substrate effects and coupling with horizontal interconnects," *IEEE Trans. Electron Devices*, vol. 58, no. 10, pp. 3529–3540, October 2011.
77. C. Gontrand, *Towards a Modeling Synthesis of Two or Three-Dimensional Circuits through Substrate Coupling and Interconnections: Noises and Parasites*, Bentham Science Publishers, Sharja, 230 pages, 2015.
78. A. He, T. Osborn, S. A. Bidstrup Alien, and P. A. Kohl, "All-copper chip-to-substrate interconnects part II. Modeling and design," *J. Electrochem. Soc.*, vol. 155, no. 4, pp. D31–D322, 2008.

79. L. Di Cioccio, P. Gueguen, F. Grossi, P. Leduc, B. Charlet, M. Assous, A. Mathewson, J. Brun, D. Henry, P. Batude, P. Coudrain, N. Sillon, L. Clavelier, G. Poupon, and M. Scannell, "3D Technologies at CEA-Leti Minatec," *4th International Conference and Exhibition on Device Packaging*, IMAPS, Scottsdale, 2008.
80. K. Banerjee, S. J. Souri, P. Kapur, and K. C. Saraswat, "3-D ICs: A novel chip design for improving deep-submicrometer interconnect performance and systems-on-chip integration," *Proc. IEEE*, vol. 89, no. 5, pp. 602–633, May 2001.
81. S. Fengyuan, "Analyse et Caractérisation des Couplages Substrat et de la Connectique dans les Circuits 3D: vers des Modèles Compacts," Connaisnces et savoirsn, 2016.
82. R. Gharpurey, *Modeling and Analysis of Substrate Coupling in Integrated Circuits. Thesis in Engineering-Electrical Engineering and Computer Sciences*, University of California, Berkeley, 1992.
83. M. N. Ali, R. Gharpurey, and R. G. Meyer, "Numerically stable green function for modeling and analysis of substrate coupling in integrated circuits," *IEEE Trans. Comput. Aided Des. Integrated Circ. Syst.*, vol. 17, no. 4, pp. 305–315, April 1998.
84. C. Christopoulos, et al., *The Transmission-Line Modeling Method: TLM*, Wiley-IEEE Press & Oxford University Press, Oxford, 1995.
85. P. S. Crovetti and F. L. Fiori, "Efficient BEM-based substrate network extraction in silicon SoCs," *Microelectron. J.*, vol. 39, pp. 1771784, 2008.
86. O. Valorge, F. Sun, J. E. Lorival, M. Abouelatta-Ebrahim, F. Calmon, and C. Gontrand, "Analytical and numerical model confrontation for transfer impedance extraction in three-dimensional radiofrequency circuits," *Circ. Syst.*, vol. 3, pp. 126–135, 2012, www.SciRP.org/journal/cs/.
87. N. K. Verghese and D. J. Allstot, "Rapid simulation of substrate coupling effects in mixed-mode ICs," *IEEE Custom Integrated Circuits Conference*, pp. 18.3.1–18.3.4, 1993.
88. A. Torabian and Y. Chow, "Simulated image method for Green's function of multilayer media," *IEEE Trans. Microw. Theory Tech.*, vol. 47, no. 9, p. 1777, 1999.
89. C. Gontrand, F. Sun, Y. Ma, J. R. Cardenas-Valdez, F. Calmon, J. C. Nuñez-Perez, and J. Verdier, "From a first order electrical analysis, towards some possible signal fluctuations consideration, for radio frequency circuits" *Microelectron. J.*, vol. 45, pp. 1061–1066, June 2014.
90. R. Gharpurey and S. Hosur, "Transform domain techniques for efficient extraction of substrate parasitics 2: Theory: The substrate green function," *ICCAD '97 Proceedings of the 1997 IEEE/ACM international conference on Computer-aided design*, pp. 461–467, 1997.
91. N. Verghese, D. J. Allstot, and S. Masui, "Rapid simulation of substrate coupling effects in mixed-mode ICs," *Proceedings of IEEE Custom Integrated Circuits Conference - CICC '93*, pp. 1–4, 1993.
92. SENTAURUSn TCAD, Synopsys Sentaurus Device. https://web.stanford.edu/class/ee328/swb/swb_menu.html.
93. Y. Ma, First order Electro-thermal compact models and noise considerations for three-dimensional integration circuits, 2018. https://hal.archives-ouvertes.fr/tel–01808972.
94. Y. Ma, L. Fakri Bouchet, F. Calmon, and C. Gontrand, "Electro-thermal modeling for three-dimensional nanoscale circuit substrates," *IEEE Trans. Compon. Packag. Manuf.*, vol. 6, no. 7, pp. 1040–1050, 2016.

95. M. Magnini, B. Pulvirenti, and J. R. Thome, "Numerical investigation of hydrodynamics and heat transfer of elongated bubbles during flow boiling in a microchannel," *Int. J. Heat Mass Transf.*, vol. 59, no. 1, pp. 451–471, 2013.
96. S. Lips, V. Sartre, F. Lefèvre, S. Khandekar, and J. Bonjour, "Overview of heat pipe studies during the period 2010–2015," *Interfacial Phenom. Heat Transfer*, vol. 4, no. 1, pp. 33–53, 2016.
97. L. U. M. R. Cnrs, "COMSOL simulation of heat pipe," *Comsol Conference*, Rotterdam, the Netherlands, 2017.
98. C. Torregiani, H. Oprins, B. Vandevelde, E. Beyne, and I. De Wolf, "Compact thermal modeling of hot spots in advanced 3D-stacked ICs," *Proc. Electron. Packag. Technol. Conf. EPTC*, no. 1, pp. 131–136, 2009.
99. W. Huang, S. Ghosh, S. Velusamy, K. Sankaranarayanan, K. Skadron, and M. R. Stan, "HotSpot: A compact thermal modeling methodology for early-stage VLSI design," *IEEE Trans. Very Large Scale Integr. Syst.*, vol. 14, no. 5, pp. 501–513, 2006.
100. W. S. Zhao, J. Zheng, S. Chen, X. Wang, and G. Wang, "Transient analysis of through-silicon vias in floating silicon substrate," *IEEE Trans. Electromagn. Compat.*, vol. 59, no. 1, pp. 207–216, 2017.
101. U. Vadakkan, S. V. Garimella, and J. Y. Murthy, "Transport in flat heat pipes at high heat fluxes from multiple discrete sources," *J. Heat Transfer*, vol. 126, no. 3, p. 347, 2004.
102. Y. Luo, B. Yu, X. Wang, and C. Li, "A novel flat micro heat pipe with a patterned glass cover," *IEEE Trans. Components, Packag. Manuf. Technol.*, vol. 6, no. 7, pp. 1053–1057, 2016.
103. L. Lin, R. Ponnappan, and J. Leland, "High performance miniature heat pipe," *Int. J. Heat Mass Transf.*, vol. 45, no. 15, pp. 3131–3142, 2002.
104. H. Li, Y. Tang, Y. Jin, B. Li, and T. Zou, "Experimental analysis and FEM simulation of antigravity loop-shaped heat pipe for radio remote unit," *IEEE Trans. Components, Packag. Manuf. Technol.*, vol. 7, pp. 1–9, 2017.
105. R. Hopkins, A. Faghri, and D. Khrustalev, "Flat miniature heat pipes with micro capillary grooves," *J. Heat Transfer*, vol. 121, no. 1, pp. 102–109, 1999.
106. G. P. Peterson, D. Wu, and B. R. Babin, "Steady-state modeling and testing of a micro heat pipe," *J. Heat Transf.*, vol. 112, pp. 595–601, August 1990.
107. M. Aghvami and A. Faghri, "Analysis of flat heat pipes with various heating and cooling configurations," *Appl. Therm. Eng.*, vol. 31, no. 14–15, pp. 2645–2655, 2011.
108. Z. J. Zuo and A. Faghri, "A network thermodynamic analysis of the heat pipe," *Int. J. Heat Mass Transf.*, vol. 41, no. 11, pp. 1473–1484, 1998.
109. U. Vadakkan, J. Y. Murthy, and S. V. Garimella, "Transient analysis of flat heat pipes," In *Proceedings of the 2003 ASME Summer Heat Transfer Conference*, vol. 3, pp. 507–517, 2003.
110. G. Patankar, J. A. Weibel, and S. V. Garimella, "A time-stepping analytical model for 3D transient vapor chamber transport," *IJHMT*, vol. 119, pp. 867–879, 2018.
111. R. S. Prasher, "A simplified conduction based modeling scheme for design sensitivity study of thermal solution utilizing heat pipe and vapor chamber technology," *J. Electron. Packag.*, vol. 125, no. 3, p. 378, 2003.
112. C. Byon, K. Choo, and S. J. Kim, "Experimental and analytical study on chip hot spot temperature," *Int. J. Heat Mass Transf.*, vol. 54, no. 9–10, pp. 2066–2072, 2011.

113. L.-M. Collin, V. Fiori, P. Coudrain, S. L. Lhostis, S. Chéramy, J.-P. Colonna, B. Mathieu, A. Souifi, and L. G., "Fréchette microchannel design study for 3D microelectronics cooling using a hybrid analytical and finite element method," *Proceedings of the ASME 2015 13th International Conference on Nanochannels, Microchannels, and Minichannels,* ICNMM2015, July 6–9, San Francisco, California, USA, 2015.
114. Y. Yadavalli, Y. Yadavalli, J. A. Weibel, and S. V. Garimella, "Performance-governing transport mechanisms for heat pipes at ultra-thin form factors performance-governing transport mechanisms for heat pipes at ultra-thin form factors," *IEEE Trans. Components, Packag. Manuf. Technol.*, vol. 5, no. 11, pp. 1618–1627, 2015.
115. R. Prieto et al., "Thermal measurements on flip-chipped system-on-chip packages with heat spreader integration," *SEMI-THERM*, pp. 221–227, 2015.
116. F. de Crécy, "A simple and approximate analytical model for the estimation of the thermal resistances in 3D stacks of integrated circuits," *International Workshop on Thermal Investigation of ICs and Systems,* pp. 1–6, September 2012.
117. W. Shockley, J. A. Copeland, and R. P. James, *Quantum Theory of Atoms, Molecules and Solid State,* Ed. P. O. Löwdin, Academic Press, New York, pp. 537–563, 1966.
118. A. Van der Ziel, *Noise in Solid State Devices and Circuits,* John Wiley & Sons, New York, 1986.
119. A. Van Der Ziel, *Noise: Sources, Characterisation, Measurement,* Prentice Hall, Englewood Cliffs, NJ, 1970.
120. K. M. Van Vliet, A. Friedmann, R. J. J. Zulstra, A. Gisolf, and A. Van Der Ziel, "Noise in single injection diodes," *J. Appl. Phys.*, vol. 46, p. 1804, 1975.
121. R. P. Jindal, "Special issue on fluctuation phenomena, in electronic and, photonic device," *IEEE Trans. Electron devices,* vol. 41, pp. 2133–2138, 1994.
122. B. Picinbono, Introduction à l'étude des signaux et phénomènes aléatoires, Dunod.
123. P. Langevin, "On the theory of Brownian motion," *Compt. Rend. Acad. Sci.*, vol. 146, p. 530–533, 1908.
124. K. K. Thornber, T. C. McGill, and M.-A. Nicolet, "Structure of the Langevin and impedance-field method of calculating noise in devices," *Solid-State Electron.*, vol. 17, p. 587–590, 1974.
125. J. P. Nougier, *III–V Microelectronics,* European Materials Research Society Monographs, vol. 2, pp. 183–238, Elsevier, New York, 1991.
126. G. Ghione and F. Filicori, "A computationally efficient unified approach to the numerical analysis of the sensitivity and noise of semiconductor devices," *IEEE Trans. Comput.-Aided Des. Integr. Circuits Syst.*, vol. 12, no. 3, pp. 425–438, 1993.
127. R. Rohrer, L. Nagel, R. Meyer, and L. Weber, "Computationally efficient electronic-circuit noise calculations," *IEEE J. Solid State Circ.*, vol. 6(4), p. 204, 1971.
128. J. B. Lee, H. S. Min, and Y. J. Park, "Reinvestigation and extension of the steady-state Nyquist theorem for multi-terminal semiconductor devices and its application to minimum noise figure in microwave field effect transistors," *J. Appl. Phys.*, vol. 79, no. 1, p. 228, 1996.
129. F. Bonani, G. Ghione, M. R. Pinto, and R. Kent Smith, "An efficient approach to noise analysis through multidimensional physic-based models," *IEEE Trans. Electron Devices,* vol. 45, no. 1, p. 261, 1998.

130. K. M. Van Vliet, A. Friedmann, R. J. J. Zulstra, A. Gisolf, and A. Van Der Ziel, "Noise in single injection diodes. I. A survey of methods," *J. Appl. Phys.*, vol. 46, no. 4, 1975.
131. J. P. Nougier, J. C. Vaissiere, and C. Gontrand, "Two-point correlations of diffusion noise sources of hot carriers in semiconductors," *Phys. Rev. Lett.*, vol. 51, no. 6, pp. 513–516, August 1983.
132. A. Papoulis and S. U. Pillai, *Probability, Random Variables, and Stochastic Process*, 3rd Edition, Mc Graw Hill, New-York, 1991.
133. A. Demir and P. Feldmann, "Modeling and simulation of the interference due to digital switching in mixed-signal ICs," *Proceedings of IEEE/ACM International Conference on Computer-Aided Design*, November, 1999.
134. H. Yasuda. "Calculation of power spectra of pulses sequenpces by means of transition matrices" *Electron Comm. Jpn.*, vol. 5A, no. 11, pp. 17–23, 1971.
135. G. L. Cariolaro, and G. P. Tronca, "Spectra of blocked coded digital signals," *IEEE Trans. Comm.*, vol. 22, pp. 1555–1563, October 1974.
136. G. Bilardi, R. Padovani, and G.L. Pierobon, "Spectral analysis of functions of Markov chains with applications," *IEEE Trans. Comm.*, COM-31, pp. 853–861, July 1983.
137. H. Kanabe, "Spectral lines of codes given as functions of finite Markov chains," *IEEE Trans. Inform. Theor.*, vol. 37, pp. 927–941, May 1991.
138. A. Demir, J. Roychowdhury, and P. Feldmann, "Modeling and simulation of noise in analog/mixed-signals communications systems," *IEEE Custom Integrated Circuits Conference*, 1999.
139. D. Ham and A. Hajimiri, "Virtual damping and Einstein relation in oscillators," *IEEE Journal of Solid State Circuits*, vol. 38, no. 3, pp. 407–418, March 2002.

Index

I

L

M

N

O

P

R

S

T

V

W